LABOI___ ____
MANUAL

F. Patterson MAIST

Former technician, universities of NSW and Manchester

ARTHUR H. STOCKWELL LTD.
Torrs Park   Ilfracombe   Devon
Established 1898
www.ahstockwell.co.uk

British Library Cataloguing-in-Publication Data.
A catalogue record for this book is available
from the British Library.

The author wishes to thank Professor K. G. Lewis
for reading the proofs and for his helpful suggestions.

By the same author:
Philosophical Studies
Memoirs of the Burma Campaign 1942-45

ISBN 0-7223-3686-1
ISBN 978-0-7223-3686-1
Printed in Great Britain by
Arthur H. Stockwell Ltd.
Torrs Park   Ilfracombe
Devon

CONTENTS

Section 4

Section 5

SECTION 1

1.1 Dangerous Chemical Mixtures

A number of chemicals only become dangerous when they are mixed, or come into contact with each other. These are known as incompatibles and a list of some of these is given below. These either form explosive mixtures, or give off poisonous fumes.

Alkali Metals — with certain liquids, particularly sodium and potassium, with water or with chlorinated hydrocarbons (CCl_4, $CHCl_3$, etc.).

Acetylene — with halogens, silver, mercury, or copper, metals or their compounds.

Ammonia — with halogens, mercury or silver.

Ammonium Nitrate — with acids, metal powders, chlorates, sulphur, and a number of inflammable liquids.

Aniline — with strong oxidising agents.

Chlorates — with ammonium salts, acids, metal powders and sulphur.

Cyanides — with acids.

Oxygen — with oils and greases, hydrogen, inflammable liquids and solids, and inflammable gases and some metal powders.

Oxalic Acid — with silver and mercury.

Potassium Permanganate — with glycerine, ethylene glycol, benzaldehyde, and sulphuric acid.

Sulphuric Acid, Conc. — with chlorates, perchlorates, permanganates, and water.

All oxidising agents, such as chromic acid, hydrogen peroxide, per acids and sodium peroxide, etc. should be regarded as potentially explosive when mixed with inflammable liquids (oxidisable substances).

A number of wall charts are available for laboratory use.

1.2 Laboratory First Aid

A first-aid box should be kept in each laboratory and/or store. It should contain the following items, and all should be clearly marked:

Bandages (1, 2 & 3 inch), lint, gauze, cotton wool, adhesive plaster and Elastoplast, Band Aids for finger cuts, a sling (triangular bandage). Forceps, scissors, safety pins. Glass or plastic eyebath, dropping tube or eye-dropper. Sal volatile, castor oil, mustard powder (half a teaspoon in water for emetic), boracic acid powder, acriflavine emulsion (min. 250 ml), saturated aq. picric acid (500 ml), boric acid 1% aq., acetic acid 1% aq., sodium bicarbonate 1% aq., iodine tincture, sodium bicarbonate saturated aq. solution (2 x 500 ml).

Solutions should be checked and renewed as necessary.

An injuries book should be kept with the box. Into it should be entered any names of persons who may have had need to use items from the box, along with the dates and times.

(See also section on safety.)

1.3 Laboratory Solutions & Reagents

Normal Solution — 1 gram/equivalent weight of substance in 1 litre of solution.
Molar Solution — 1 gram/molecule of substance in 1 litre of solution.
Acids — Concentrated acids are: Nitric — 16N; Sulphuric — 36N; Hydrochloric — 10N; Glacial acetic — 17N.

Many of these reagents are used in mixed (organic & inorganic) laboratories.

Dilute Acids — HNO_3 (2N) dilute 250 ml with distilled water, to 2 litres. H_2SO_4 (2N) dilute 110 ml conc. (1.848 SG) to 2 litres, with water. Always add acid to water with care!

To obtain 20% sulphuric acid, use 270 ml to 1 litre water.

HCl (2N) dilute 400 ml conc. (1.1685 SG) to 2 litres H_2O (use distilled water).

HAc — acetic — (2N) dilute 232 ml glacial acetic acid, to 2 litres with distilled water.

Acetylating Mixture

Equal volumes of acetic anhydride and glacial acetic acid.

Ammonia, Dilute Solution (2N)

200 ml 0.88 SG liquor, to 2 litres (use distilled water).
250 ml 0.89 SG liquor, to 2 litres (use distilled water).
(For 4N, sometimes preferred, use double quantities of liquor.)

Ammoniacal Silver Nitrate

2-3 drops of aq. NaOH to about 5 cc silver nitrate solution, in a t.t., then add dilute NH_4OH drop by drop until a trace of Ag_2O remains. Tollen's Reagent has now replaced this for safety reasons (see later).

Alcoholic KOh

5 g KOH, made up to 100 ml with ethanol.

Alternative method:
10 g KOH, with 100 cc ethanol. Reflux 30 mins, cool, decant and filter.

Barium Chloride

$(BaCl_2.2H_2O)$ 122.15 g/litre = 1N.

Bromine in Carbon Tetrachloride

0.75 ml Br_2 in 100 ml CCl_4.

Bromine in Water

Saturated sol. Shake a few ml Br with water until no more dissolves (fume cupboard).

Calcium Chloride

$(CaCl_2.6H_2O)$ 109.55 g/litre = 1N.

Chlorine Water

Saturated solution Cl_2. Saturate with Cl_2 from cylinder, or HCl (conc.) on $KMnO_4$ in chlorine generator.

Chromic Acid

12 g $K_2Cr_2O_7$ (or Na) made up to 250 ml with conc. sulphuric acid.

Congo Red

1 g in 1 litre of water.

Deniges Test (for Acetone)

2 ml 1% sol. acetone in water, add 2 ml acid mercuric sulphate reagent. Place in beaker of hot water. While ppt formed due to $HgSO_4$ + acetone.

Dichromate Oxidising Mixture

20 g dichromate ($Na_2Cr_2O_7$ or $K_2Cr_2O_7$).
100 ml H_2O.
8 ml H_2SO_4.

DNP Reagent

1) Dissolve 2 g dinitrophenylhydrazine in 15 ml conc. H_2SO_4. Add 150 ml methanol, and dilute to 500 ml with water. Allow to cool, and filter.

Alternative:

2) Dissolve 0.258 g DNP in a mixture of 42 ml conc. HCl, add 50

ml of water (warm on bath to dissolve). Cool, and dilute cold solution with distilled water to 250 ml.

Distillation of Ether Residues

Before distilling, shake with aqueous ferrous sulphate to reduce the risk of an explosion.

Alternatively, distil from a piece of shiny copper foil or wire.

Fehlings Solution "A"

35 g pure $CuSO_4$ in 500 ml distilled water.

Fehlings Solution "B"

70 g NaOH + 175 g Rochelle salt in 500 ml distilled water.

Ferric Chloride

1) Laboratory Reagent — 1% w/v acidify with 1 ml of conc. hydrochloric acid.

2) Neutral = 90 g/litre (1N) add NaOH until precipitate just formed, then filter.

Ferric Chloride

Approx. 4.5%.

Dissolve 75 cc $FeCl_3.6H_2O$ in water, add 10 cc conc. HCl, make up to 1 litre.

$FeCl_3$ (anhydrous) MW = 162.

$FeCl_3.6H_2O$ (hydrate) MW = 270.3.
If using hydrate 135 g/litre = N/2.

Hydrogen Peroxide

20 vols. = 3.6N = 6.08% dilute 100 ml of 100 vols. strength to 500 ml for 20 vols.

Hydroxylamine Hydrochloride

For *aldehydes and ketones* 25% aq.

Hydroxylamine Hydrochloride

For *esters and anhydrides* 7 g KOH pellets in 12.5 ml water add 50 ml methanol.

Hydroxylamine Test for Esters and Anhydrides

5 to 7% methanolic sol. of hydroxylamine hydrochloride used. Heat mixture of ester and above, and methanolic KOH. Acidify (HCl) add drops of ferric chloride sol. (violet colour).

Iodine

For iodoform reactions, 2 g iodine — 10 g KI, in 30 ml water.
0.1N solution is 12.7 g in solution of 20 g KI, in 30 ml water, to 1 litre of water.

Lead Acetate

95 g lead acetate in 1000 ml water, add acetic acid to clear solution (= 0.5N).

Litmus

Azolitmin 0.05 g/100 ml water (if using extract, adjust colour with NaOH and HNO_3).

Lime Water

Approximately 10 g CaO per litre. Filter or stand and decant.

Methanolic Potassium Hydroxide

Prepare by dissolving 7 g KOH pellets in 12.5 ml water then methanol solution approx. 2N in 80% methyl alcohol.

Methyl Orange

1 g Methyl Orange dissolved in ethanol (about 20 ml) made up to l litre with 50/50 ethanol/H_2O.

Methyl Orange (screened)

Dissolve methylene blue (1 g approx.) in ethanol, add this drop by drop to M.O. solution until red with acid, and green with alkali.

Methyl Red

lg in 600 ml alcohol. Dilute with 400 ml water (or use aliquot amounts).

Nessler's Reagent

100 ml 5% mercuric chloride, add 5% KI until precipitate just re-dissolves, then make alkaline with NaOH (yellow colour with traces of NH_3).

Sodium Bisulphite Solutions

Alternative. Pass SO_2 into a solution of 10% NaOH (until no longer alkaline to litmus). ($NaOH + H_2SO_3 = NaHSO_3 + H_2O$) 5.2% $NaHSO_3$ solution (see later methods).

Sodium Carbonate

(Na_2CO_3) 53 g/litre = 1N. (aq.)
$10H_2O$ 143.08g/litre = 1N.

Sodium Hydroxide

40.01 g/litre = 1N.

Sodium Thiosulphate

$(Na_2SO_3.5 H_2O)$ 248.21 g/litre = 1N.

Sodium Hypobromite

200 g NaOH in water, make up to 1l. Chill in fridge, add 50 cc bromine, stirring (fume cupboard).

Sodium Hypobromite

Dissolve 200 g NaOH in distilled water (750 ml). Chill in fridge. Add 50 ml bromine to solution in 3 necked flasks from a sep. funnel. Add drop at a time and keep the temp. as low as possible (10°C). Vol. should not exceed 1200 ml. Store in fridge. Gives 10% solution.

Sodium Metaperiodate

0.1M. 21.4 g/litre (accurate).

Starch Solution

Make a paste with 1 g starch and 10 ml cold water. Add slowly stirring to 90 ml boiling water. Boil for 5 mins. Add a few drops of chloroform. (May also be preserved for a long period by keeping under a layer of toluene in a stoppered bottle.)

Silver Nitrate

169.89 g/litre = 1N.
Lab reagent 5% w/v aq.
Lab reagent 2.5% w/v alcoholic.
(For chloride tests 0.05N solution.)
Use distilled water.

Tollen's Reagent

Dissolve 10 g silver nitrate in 100 ml water (sol. A).
Dissolve 10 g sodium hydroxide in 100 ml water (sol. B).
(For testing add v. dil. NH_4OH in drops until ppt just dissolves) = silver mirror with aldehydes on warming in hot water. (When dry can be explosive.)

Phenolphthalein

0.05 g in 100 ml 5% alcohol.

Potassium Iodide

N/10 solution = 16.6 g/litre.

Phenyl Hydrazine in Acetic Acid

25 ml (pure) *distilled phenyl hydrazine* (137°C/18 mm pressure) in 250 ml 10% acetic acid. Add 0.5 g charcoal, shake and filter.

Phosphate Buffer

39 g di-sodium hydrogen orthophosphate + 13.6 g. Potassium dihydrogen orthophosphate. Dissolve in water to make 2 litres, gives a pH of 6.9 (M/10) for periodate determination.

Schiff's Reagent

Dissolve 0.2 g rosaniline (or one of its salts) in 10 ml freshly prepared solution of SO_2 aq. Allow to stand until pale red disappears, and it becomes colourless or pale yellow. Dilute with water to 20 ml (use stoppered bottles).

Sodium Arsenite

1.25 g As_2O_3 to 10 cc NaOH (2N), warm to dissolve. Acidify with HCl dil., add 100 ml cold saturated $NaHCO_3$ sol. — make up to 250 ml (in solution).
Molar sol. contains 197.82 g/litre.
0.015M = 2.967 approx. 3 g/litre. (5.935 g per 2 litres.)
Dissolve in minimum amount NaOH.

Sodium Bisulphite Reagent (Fume cupboard)

SO_2 passed into washing soda ($Na_2CO_3.1OH_2O$) crystals covered with a layer of water until an apple green sol. is obtained. Or use 60% metabisulphite sol., pass SO_2 a few mins. to ensure absence of sodium sulphite.

Sodium Bisulphite (Alcoholic)

Saturated sol. of sodium bisulphite aq., add 70% volume of alcohol, then just sufficient water to produce a clear solution.

Inorganic Reagents Used in Elementary Chemical Analysis

Distilled water should be used wherever possible.
Acids — as in the organic section.

Alumin on Reagent (Ammonium salt of aurine tricarboxylic acid)

1 g/litre aq.

Ammonium Acetate Solution

154 g/litre aq. 2M.

Ammonium Carbonate Solution

$(NH_4)_2CO_3$ 200 g/800 ml aq. and 200 ml of 0.89 SG. ammonia solution 2M.

Ammonium Chloride

20% aq., or 25% NH_4Cl in ethanol. Solid may be used instead of solution.

Ammonium Hydroxide

Approx. 4N 270 ml of 0.88 SG NH_4OH per 1 litre (aq. dist.).

Ammonium Molybdate

40 g molybdenum trioxide in 70 ml of NH_4OH (0.88 SG) and 140 ml water. Add while stirring, 250 ml conc. HNO_3 and 500 ml dist. water. Dilute to 1 litre. Decant if necessary.

Ammonium Oxalate

Saturated solution aq.

Yellow Ammonium Sulphide (for b. metals of group 2)

150 ml conc. NH_4OH saturate with H_2S and cool. 250 ml conc. NH_4OH add 10 g of powdered sulphur. Shake to dissolve, and dilute to one litre aq.

Ferrous Ammonium Sulphate

0.5N.
196 g $Fe(NH_4SO_4)_2.6H_2O$ in 1 litre of water, containing 10 ml of conc. sulphuric acid.

Hydrogen Sulphide

H_2S. Saturated solution in acetone, renew every 3 weeks. Alternatively, 30 g/litre aq. thioacetamide (for group 2).

8-Hydroxyquinoline Reagent

5 g in 100 ml of alcohol.

Lead Nitrate

Saturated aq. solution.

Lead Acetate

95 g $Pb(C_2H_3O_3)_2.3H_2O$ in 1 litre of water. Add acetic acid to clear.

Lithium Hydroxide Reagent

0.4N.
lg lithium hydroxide (or 1.75 g $LiOH.H_2O$) in 50 g of potassium nitrate, in 100 ml of water. Use a soda lime guard tube.

Magnesia Mixture

55 g $MgCl_2.2H_2O$ added to 135 g NH_4Cl dissolved in water, add 350 ml of 0.88 SG ammonia, and dilute to 1 litre aq. dist.

Magneson 1

Dissolve 0.25 g p-nitrobenzeneazo-resorcinol in 1 litre N/1 sodium hydroxide solution.

Magnesium Uranyl Acetate

100 g uranyl acetate in 60 ml glacial acetic acid, dilute to 500 ml aq. Dissolve 330 g magnesium acetate in 60 ml glacial acetic acid, dilute to 200 ml. Heat to boiling, until clear. Pour the magnesia solution into the uranyl acetate solution. Cool, and dilute to 1 litre aq.

Mercuric Chloride

27 g $HgCl_2$ in 1 litre aq. 0.2N.

Molisch's Reagent

15 g naphthol in 100 ml alcohol (or chloroform).

Millon's Reagent

One part mercury (by wt.) to 2 parts conc. nitric acid. Dilute twice, aq., and decant if necessary.

Nessler's Reagent

50 g KI in 50 ml aq. Add mercuric chloride, 22 g in 350 ml aq. Add 200 ml of 5N NaOH, dilute to 1 litre and decant if necessary (NH_3 yellow colour).

Nitron

10 g diphenylendo-anilo-dihydro-triazole, in 100 ml 5% acetic acid.

α Nitroso β Naphthol

1 g in 50 ml glacial acetic acid. Dilute to 100 ml aq.

Potassium Dichromate (or bichromate)

2% aq.

Potassium Ferricyanide

55 g/litre aq. 0.5N.

Potassium Ferricyanide

53 g/litre aq. 0.5N.

Potassium Iodide

2% aq. w/v.

Potassium Perchlorate

5% aq. w/v.

Rhodamine B

0.01% aq.

Schweitzer's Solution

Add 0.88 ammonia to concentrated copper sulphate solution until precipitate dissolves. (Solvent for cellulose.)

Sodium Cobaltinitrite

17 g $Na_3(Co(NO_2)_6$ in 250 ml aq.

Stannous Chloride

56 g $SnCl.H_2O$ in 100 ml conc. hydrochloric acid, dil. to 1 litre aq. Metallic tin in the bottle helps to preserve.

Thiourea

10 g in 100 ml aq.

Titan Yellow

0.1% aq.

Zinc Uranyl Acetate

20 g uranyl acetate to 100 ml aq. add 4 ml acetic acid, warm, add 60 g cryst. zinc acetate in 2 ml acetic acid gl. Add 100 ml water.

Zirconium Nitrate

100 g zirconium nitrate in 1 litre N/1 nitric acid. Decant (for phosphate separations).

Indicators

(Some of these vary slightly from those in the organic section, both work equally well.)

Congo red	1 g in 1 litre aq.
Dichlorofluorescein	0.1 g in 100 ml 70% alcohol.
Eosin	0.1 g in 100 ml 70% alcohol.
	(or sodium salt of eosin may be used).
Fluorescein	0.1 g in 100 ml 70% alcohol.
Methyl orange	1 g in 1 litre of 70% alcohol.
Methyl red	1 g in 1 litre of 70% alcohol.
Phenolphthalein	1 g in 100 ml of 50% alcohol.

Suitable Drying Agents

For drying alcohols,	Anhydrous K_2CO_3, $MgSO_4$, $CaSO_4$, CaO (quicklime).
For drying alkyl halides,	Anhydrous $CaCl_2$, Na_2SO_4, $MgSO_4$, $CaSO_4$, P_2O_5.
For drying aromatic hydrocarbons	Anhydrous $CaCl_2$, $CaSO_4$, Na (metal), P_2O_5.
For drying ethers	Anhydrous $CaCl_2$, $CaSO_4$, Na (metal), P_2O_5.
Aldehydes	Anhydrous Na_2SO_4, $MgSO_4$, $CaSO_4$.
Ketones	Anhydrous Na_2SO_4, $MgSO_4$, $CaSO_4$, K_2SO_4.

1.4 Universal Indicators

Phenolphthalein — 0.08 g.

Naphthophthalein — 0.18 g.

Methyl orange — 0.04 g.

Methyl red — 0.02 g.

Dissolve the above in 100ccs of 70% alcohol and distilled water.

Colour of Caustic Solution with Universal Indicator

pH	Cc N/5 NaOH per 100 cc of buffer solution in 200 cc	Colour with Universal Indicator
2.0	0	-
3.0	20	crimson
4.0	28	red
5.0	36	orange-red
5.5	40	orange
6.0	45	orange-yellow
6.5	48	yellow
7.0	52	greenish-yellow
7.5	56	yellowish-green
8.0	60	green
8.5	65	bluish-green
9.0	70	greenish-blue
9.5	73	blue
10.0	77	violet
11.0	85	reddish-violet
12.0	100	-

1.5 Purification of Solvents for Research

Purification of Ethyl Acetate

Reflux the crude solvent with acetic anhydride (pretreated with a few mls of concentrated sulphuric acid) in the proportions of 10/1, for 4 hours. This treatment will convert any water present to acetic acid, and any ethanol to ethyl acetate and acetic acid. Although there will be an excess of anhydride, the widely separated b.p.s of this, the acetate and acetic acid enable separation to be carried out by fractional distillation, the range of acetate to be retained being 78°C $\pm$ 1 (760 mm Hg). Then dry over K_2CO_3 followed by re-distillation.

Purification of Methanol

Reflex crude solvent with orthophosphoric acid (H_3PO_4) 88/99.3% (125/1) for one hour, then distil onto potassium hydroxide pellets. Decant and re-distil, retaining the fraction at 65°C. Simple distillation produces a satisfactory solvent for most extraction purposes.

Purification of Acetone for Research Purposes

Reflux crude Me_2CO with powdered $KMnO_4$ add in portions until the refluxing action no longer occurs any colour change. Distil onto anhydrous potassium carbonate, filter and re-distil using a short fractionating column, retaining purified solvent of boiling range 56°C. A water bath is recommended for the reflux stage, and care must be taken not to contaminate the electrothermal mantle.

For general purposes acetone may be purified by standing over anhydrous potassium carbonate for about 12 hours, then filtering and distilling.

Preparation of Ethanol for General Use

Distil the crude product using a Kjeldahl splash head, and retain the fraction at 78°C (760 mm Hg).

Purification of Solvents for Infrared Spectroscopy

These methods are suitable for small amounts, where reference samples are required for infrared spectroscopy.

The product (50 cc) is passed through a glass column (1 cm diameter) containing about 3 inches of chromatographic alumina, (Al_2O_3). The liquid is then placed over a molecular sieve for a minimum of one hour, and then distilled into clean, dry, glass receptacle. The initial and final distillates are discarded, and only a 'middle range' is taken. This method is suitable for hexane, cyclohexene, octene, etc.

Purification of Acetic Acid for Use as a Hydrogenation Solvent

250 ml of acetic acid in a round-bottomed flask, 1.3 g $KMnO_4$ are added. The side arm is blocked, and held at an angle to drain back into the flask, the acid is refluxed for one hour. Then distil into a bottle, which should previously have been washed with warm NaOH solution and then distilled water. The initial and final fractions are rejected. The rubber stopper used should be boiled for 30 mins in 10% NaOH, rinsed in distilled water, and dried in air.

Purification of Benzene for Research Purposes

Reflux crude solvent with mercuric acetate (30/1) for 4 hours, following this by azeotropic distillation, using interchangeable condensers. If the distillate appears to contain water droplets, this should be dried over $CaCl_2$ for 12 hours, followed by re-distillation (see over for thiophene determination).

Alternatively, crude benzene maybe agitated with concentrated H_2SO_4 (10/1) repeating until the acid is colourless. The compound thiophene dimercurhydroxyacetate is formed, which gives mercuric chloride on distillation (from this the acetate can be prepared). The isatin reaction can be used to determine the presence of thiophene. For this reaction, 10 mg ∓ isatin is dissolved in 1 ml of conc. H_2SO_4 and this when shaken with the sample gives a blue/green colour.

Purification of benzene, for general purposes requires distillation through an efficient fractionating column, which is usually lagged and packed with Fenske helices (b.p. 78°C-80°C).

Purification of Diethyl Ether

To minimise the danger of explosion, and/or later polymerisation, peroxides and hydroxides should be removed initially by agitating crude ether with a concentrated solution of ferrous sulphate (60 g $FeSO_4$ 6 ml conc. H_2SO_4 and 110 ml aq. dist.) followed by separation and repeated washings with water. Test the solvent for peroxides and if negative, dry the ether over $CaCl_2$ (which also removes any ethanol) for 12 hours. Filter and distil, using an efficient condenser and column (preferably a 'closed' arrangement with a tube from a vacuum take-off to the outside air (b.p. 35°C).

Sodium-dried ether may be prepared by treating the above solvent with two successive pressings of fine-gauge sodium wire, taking care to insert a drying tube in the stopper of the store bottle.

Purification of Ethanol for Spectroscopy (IR & UV)

Reflux ethanol in a round-bottomed flask with KOH pellets and zinc dust (50:1:1) for two hours; then distil off the fraction boiling at 78°C (760 mm Hg). Use an efficient fractionating column (Vigreux) and ensure absence of splashing by regulation of heat input.

Preparation of Super-Dry Ethanol

Dry a 1.5 or 2 litre round-bottomed flask and double surface condenser, with drying tube. Add 5 g of magnesium turnings, and 0.5 g of iodine, then 50/75 ml of 99% ethanol. Warm until all the Mg dissolves then add 900 ml of ordinary ethanol. Reflux 2 hours then distil excluding air (78°C, 760 mm Hg).

1.6 Correction of Boiling-Point Temperatures at Various Pressures

Temperature correction — 0.00012(760 - p) (273 - t).

Temperature where p = actual barometric pressure, in mm of mercury.

Temperature where t = observed boiling point.

The Correction of Barometric Readings at Various Latitudes

In latitudes from 0 to 45 the correction is added, and deducted for latitudes from 45 to 90.

Latitude	Height of Barometer in mm of Mercury						
	680	700	720	740	760	780	
25	1.14	1.17	1.20	1.23	1.27	1.30	add
30	0.89	0.91	0.93	0.96	0.98	1.01	add
35	0.60	0.62	0.64	0.66	0.67	0.69	add
40	0.30	0.31	0.32	0.33	0.34	0.35	add
50	0.30	0.31	0.32	0.33	0.34	0.35	subtract
55	0.60	0.62	0.64	0.66	0.67	0.69	subtract
60	0.89	0.91	0.93	0.96	0.98	1.01	subtract
65	1.14	1.17	1.20	1.23	1.27	1.30	subtract

1.7 Filter Papers

Whatman		*Porosity*	
1		medium	rapid
2		medium	medium
3		fine	medium
4		coarse	rapid
5		fine	slow
30		medium	rapid high wet strength
31		coarse	rapid high wet strength
40	ash-free	medium	medium
41	ash-free	coarse	rapid
42	ash-free	coarse	rapid thin hard
43	ash-free	fine	slow fine ppt
44	ash-free	fine	slow for very fine ppt
50		fine	slow for caustic sol.
52		medium	medium
54		coarse	rapid

1.8 Preparation of Chromatographic Aluminas

The type of alumina most suitable for a given application cannot be predicted with certainty unless similar uses have been previously recorded but the following examples of chromatographic separations will serve to illustrate the kind of considerations that should govern a choice.

The 'standard' weakly basic alumina (Brockmann Activity 1, see later), is particularly useful for separating essentially neutral organic compounds (eg. hydrocarbons). It is also very effective for solvent drying and purification and has quite marked cation exchange properties. Neutral alumina is often preferable when dealing with absorbates that are pH sensitive and this grade shows weak cation exchange properties. On the other hand the weakly acidic alumina (pH 6) is especially useful for the chromatography of amino acids (amphoteric substances) and the more strongly acidic grade (pH 4.5) has marked anion exchange properties and provides an effective means of separating mixtures of strongly acidic organic substances such as carboxylic and sulphonic acids.

Hydration and Activity

The Brockmann classification of activity is assessed according to the absorptive power shown towards a series of azo-dyes is essentially a consequence of the degree of hydration of the active centres. Activity 1 represents the highest activity (greatest adsorptive power) and the following table shows the relationship between these and their corresponding water contents:

Brockmann Activity	Water Addition (%)
1	0
2	3
3	6
4	10
5	15

Reaction of a 5% aqueous suspension, measured electrometrically after 10 minutes equilibration.

Preparation of Chromatographic Aluminas of Definite Graded Adsorptive Capacity

The designations Brockmann 1, 2, 3, 4 and 5, in order of decreasing activity and alumina may be adjusted to the required grade by variation of its water content. In the case of most chromatographic aluminas, this is accomplished simply and accurately by isothermal exposure of thin layers of the alumina, contained in laboratory desiccators, to atmospheres of constant relative humidity (R.H.). Such atmospheres may be obtained by charging the desiccators with saturated solutions of certain salts, some of which are shown in table A. This simple technique has the advantage in that successive batches have precisely the same activity within the range of the particular Brockmann grade chosen.

Table A

Solutions of		Formula	% R.H. of Atmosphere at 20°C
Zinc Chloride		$ZnCl_2.3/2H_2O$	10
Potassium Acetate		$CH_3.COOK$	20
Calcium Chloride		$CaCl_2.6H_2O$	32.3
Zinc Nitrate		$Zn(NO_3)_2.6H_2O$	42.0
Sodium Dichromate		$Na_2Cr_2O_7 2H_2O$	52.0
Sodium Bromide		$NaBr.2H_2O$	58.0
Potassium Nitrate	} Mixed saturated	KNO_3	72.6
Ammonium Chloride	solutions	NH_4Cl	
Sodium Phosphate		$Na_2HPO.12H_2O$	95.0

Preparation of a Neutral Alumina from Standard Chromatographic Alumina

For some applications, a neutral chromatographic alumina is essential and this may be prepared in the laboratory from standard types of alumina, which are *slightly alkaline*.

Slurry the alumina with 2% ammonium hydroxide at 70-80°C. Stir for approximately 30 minutes and decant the wash liquor.

Continue water washing by decantation until chloride free, giving a final wash using dilute acetic acid (pH 6.5 approx.).

Remove free water on a Buchner funnel, oven dry at 120°C, and calcine for 3 hours at 600°C.

The methods given for preparing and recovering aluminas, are suitable for plate or column chromatography.

Spreaders are available for giving standard-thickness alumina. Prepared acetylamide sheets are also available. They are discarded after use.

Recovery of Chromatographic Alumina

Wash with ethanol — Dry (air, then oven 120°C).
Wash with dil. acetic acid (approx. 2N) — Filter 75 - 100 sieve.
Wash with dil. ammonia — Heat 600°C for 3 hours.
Wash with distilled water — This is now ready to use.

See data sheet for Brockmann Activity.

1.9 Ion-Exchange Resins

Amberlite 50 — Weakly acidic ion-exchange resin. Cross-linked polystyrene, unifunctional.

Amberlite 120 — Strongly acidic. Unifunctional.

Amberlite 400 — Strongly basic. Unifunctional.

Amberlite 410 — Strongly basic. Unifunctional.

Note: All the above resins can be regenerated (methods are given below).

Permutit — Biodeminerolit, mixed resin.

The following table gives the equivalent ion-exchange resins, made by four different companies:

Permutit (London)	Rohm & Haas (USA)	Dow Chemical Co. (USA)	ChemicalProcess (USA)
z|eo - Karb 215	Amberlite IR - 1	Dowex 30	Duolite C - 10
z|eo - Karb 225	Amberlite IR - 120	Dowex 50	Duolite C - 20
z|eo - Karb 226	Amberlite IRC - 50		
De - Acidite E	Amberlite IR - 4B	Nalcite WBR	Duolite A2
De - Acidite F	Amberlite IRA - 400	Dowex 1	
	Amberlite IRA - 410	Dowex 2	
De - Acidite G	Amberlite IR - 45		

Regeneration of Ion-Exchange Resins

Technique

Backwash — use deionised water.
Regeneration — about 1.0 normal (1N).

Rinse — deionised water. Where exhaustion involves other solvent, displace water with solvent; if not water-miscible use mutually miscible solvent as intermediate rinse.

Volumes of Solutions

Exhaustion Step — for *quantitative* exchange use amount which will utilise less than half of the ion-exchange capacity of the resin as measured in the column.

Backwash — use as much as is necessary to achieve required results.

Regeneration — use amounts calculated to provide from 150 to 500% in excess of theoretical capacity of resin in column.

Rinse — use enough to rinse excess regenerant from the column. Usually 10 bed volumes of deionised water are adequate. Rinse can be checked chemically to ensure completeness. About 40 to 50% of the total volume occupied by an ion-exchange resin in a column is void space. The exhausting solution will displace the water or solvent filling this space so it is usually desirable to discard the first bed volume before collecting the treated solution *unless* a quantitative experiment is being performed.

Method

To reuse an ion-exchange resin after an exhaustion step, or to convert it to another ionic form for a different experiment, a proper regeneration technique must be used. Column operation allows for a high level of regeneration and the same column used in the exhaustion regenerant is applied and adequate rinsing should always follow application of the regenerant. Suitable flow rates are given below. Note that the first bed volume of rinse water should be applied at the regeneration flow rate since this actually represents the completion of regeneration. In general, concentrations of about 1N are used, or from about 4 to 10%. The regeneration step requires 0.08 to 1 bed volume per minute. The method of calculation is given in the next paragraph. Applied at the suggested flow rates, such concentrations will allow for adequate contact time if the correct volumes of regenerants are used.

The correct volume of a regenerant depends upon the ion-exchange resin in question. It is convenient to express such volume in terms of bed volumes of resin. This can be calculated from the following:

$$\text{Bed Volumes of Regenerant} = \frac{\text{Vol. Capacity of Resin} \times \text{Regenerant Requirement}}{\text{Normality of Regenerant}}$$

The 'regenerant requirement' can be determined from the table for various resin types and ionic forms. The volume capacity of the particular ion-exchange resin must be expressed as milli-equivalents per millilitre of resin or some other equivalent term.

Suggested Regeneration Levels

Ion-Exchange Resin	Ionic Form	Regenerant	Requirement (Meq. Resin)	Example
Strong Acid		(HCl)	3 - 5	1R 120H
Cation	H+	(H_2SO_4)		
	Na+		3 - 5	
		NaCl		1RC 50(H)
Weak Acid		(HCl)		
Cation	H+	(H_2SO_4)	1.5 - 2	
	Na+	NaOH	1.5 - 2	
Strong Base	OH-	NaOH	4 - 5	1RA 400(C)
Action 1		(NaCl)		
	Cl-	(HCl)	4 - 5	1RA 401(C)
		(Na_2SO_4)		
	SO_4=	(H_2SO_4)	4 - 5	
Strong Base	OH-	NaOH	3 - 4	
Action 2				
Weak Base		(NaOH)		
Anion	Free Base	(NH_4OH)	1.5 - 2	1R 45 (OH)
		(Na_2CO_3)		
	Cl-	HCl	1.5 - 2	
	SO_4=	(H_2SO_4)	1.5 - 2	

The strength of the regenerant used is not critical, but normal is recommended. The limits should be between 4 and 10%. A standard technique, however, should be adhered to. This is necessary for the purposes of flow rate and timing.

Laboratory work of an analytical nature requires as high a degree of regeneration as possible. Levels of regeneration calculated in the above manner are designed to ensure that regeneration will accomplish this.

37

The flow rate of a column during ion exchang, should be 0.1 to 0.2 (i.e. about 1/10th vol./min.) bed volumes per minute.

In all cases of regeneration it would be necessary to check the quality of the water by conductivity measurements, to ensure that the process had been efficient. This would also apply to other liquids which were purified with ion-exchange resins.

SECTION 2

2.1 Radiation and Radio-Biological Safety Measures

The safety precautions that apply to chemical and biological laboratories also apply to laboratories which are concerned with radioactive experiments, or areas in and around atomic piles.

Radiation damage to animals, or human tissue, may occur from one of two sources, or both. These are:

(a) Internal tissue radiation, due to ingestion of radioactive particles such as dust. This may be α (alpha) or β (beta) radiation.

(b) External radiation, usually due to γ (gamma) rays. X-rays produce similar effects.

The amount of damage done by these processes is due to the production of ions (or ionisation), which is here a form of disintegration of living tissue.

A similar process can be produced by neutron bombardment of tissue. A fast neutron would do less damage than a slow one, since it would have less chance of striking other atoms, and therefore less ionisation would result.

Measurement of Radiation

Radiation may be measured by: (a) Geiger counters.
(b) Electroscopes.
(c) Photographic films.

Electroscopes measure radiation by determining to what extent air has become conducting as a result of ionisation (produced by radiation). In the case of radioactive dusts air sampling must be performed, and a rise in measured radiation then compared with standards. Similar modified methods may be used for liquids.

Biological effects resulting from radiation cannot easily be distinguished from those produced by other chemicals, or, in some cases, by disease (unless radiation is known to exist).

Radiation

Observations were made before 1900 by French physicians, who found certain symptoms amongst radium workers. They noted pulse and heart irregularities, headaches, nausea, and vomiting. Later research showed that radiation produced an increased cell permeability to ions.

The Main Body Changes in Acute Radiation

1 — Increase in sedimentation time and clotting time, that is delayed coagulation.

2 — Increase in capillary fragility and cardiovascular damage. Anaemias appear after about two days.

There is also a loss of mineral chlorides. In the case of ingestion, heavy metals deposit in the bones (Pb, Hg, As, Ra, Y, Pu). The rate of elimination is of the order of 0.005% per day.

Biological Effects of Radiation

Small amounts of radiation do not appear to have any harmful effects. The atmosphere contains traces of radioactive gases, which, along with cosmic rays, constitute natural radioactivity. There is a slight increase in radiation, due to cosmic rays, as one increases in altitude. These variations are shown below:

Altitude	Total Radiation in Thousandths of a roentgen per 24-Hour Period (Due to Natural Radioactivity)
10 miles high, in an aircraft.	About 90
Mount Everest, 29,002 ft	36
Mont Blanc, 15,800 ft	18
Mount Kosciusko, 7,327 ft	6.2
1,000 ft	2
Sea level	0.75

Radiation damage, or *toxic effects*, of radioactive substances in the body depend upon the following factors:

1 — Radioactive half-life of the substances.

2 — Energy and type of radiation.

3 — Location in the body, or site, and/or selectivity.

4 — Amount of radioactive substance involved, and its rate of elimination.

Gross overexposure to radiation can produce, in humans, radiation sickness, nausea, vomiting, diarrhoea, bleeding and blood changes. Death may result in untreated cases.

Smaller amounts of radiation may be indicated by changes in the blood and blood-forming organs. These effects usually manifest themselves from two to ten days after exposure. Other long-term effects may result.

Changes in the blood can indicate, or not, a person working in a radioactive area has received an excessive amount of radiation.

Blood counts performed periodically (say every two months) would safeguard workers in radioactive areas. The comparison with normal blood standards would indicate overexposure, or otherwise. If overexposed, then the removal to non-radioactive areas would ensure that recovery would be possible.

2.2 Safeguarding Workers from Radiation

As mentioned previously, regulations which apply to most chemical laboratories also apply to radioactive laboratories, and other areas where there is radioactivity.

Additional Rules for Protection from Radiation

1 — Always wear film badges, or personal ionisation chambers, when in a radioactive area. Be sure that these are checked at regular intervals.

2 — Wear protective clothing where necessary. (Specific instructions should apply to certain areas, and should specify types of clothing to be worn.)

3 — Use glove boxes, or other means, when handling or transferring radioactive materials.

4 — Restrict all radioactive experiments to given areas, and ensure, by notices, that other workers know which areas these are.

5 — Monitor areas used for radioactive work frequently, and take air samples.

6 — Perform regular haematological tests on workers in radioactive areas.

Disposal of Radioactive Waste

There are a number of problems associated with radioactivity. These include the disposal of waste, particularly if it is material of long half-life. Wherever possible, these should be marked and kept in closed containers until the material becomes less radioactive, or until burial is possible. Disused mines, or areas allocated for disposal, appear to be the only final means of disposal.

Decontamination

Decontamination of instruments may be necessary on some occasions. Wherever possible, in the case of clothing, disposal would probably be better than attempting decontamination. With regard to the surface absorption, which would be likely to occur (stainless steel for example would be easier to decontaminate than wood), the surface may be washed by appropriate chemical solutions and/ or solvents. Scraping, in the case of bench tops for example, followed by a number of chemical washings and water, would probably reduce the contamination. A means of collecting the liquids used in washing or a means of suction would be necessary.

The long-term contamination of vegetation by radioactive isotopes is in itself an environmental problem. The only answer at present appears to be in the selection of suitable sites.

A solution of 0.5N hydrochloric acid is a suitable solvent for washing metal surfaces in the decontamination process.

2.3 The Roentgen, Curie and Maximum Permissible Levels

Roentgen

= 1 electrostatic unit (1 esu) of ionising electricity in one cubic centimetre of air, at normal temperature and pressure (NTP).

= 1.6×10^{12} ions per gram of air (1cc - 0.001293 g). (These are ion pairs.)

Sievert

The sievert (sv) is a unit of dose equivalent (joules/kilogram).

The Microcurie

The microcurie is the amount of radioactivity which exists if there are 3.7×10^4 disintegrations per second (represented thus: μc). The curie is defined as the amount of radioactive material which decays at the rate of 3.7×10^{10} disintegration per second. One disintegration per second is a becquerel.

Radiation Standards

Background, due to cosmic radiation at sea level, may be taken as 0.0001 roentgens per 24 hours. Variation is due to small amounts of radioactive gases present.

The External Permissible Level

The external permissible level for radiation in any form is 3 roentgens for a 13-week period.

One milligram of radium at 1 cm distance gives an intensity of radiation of 64 roentgens in 8 hours. This is approximately 8.4 roentgens per hour at 1 cm distance.

The inverse square law applies (intensity α 1/distance^2).

A total of 500 roentgens (for total body radiation) is considered fatal in 50% of cases.

Air Sampling

The rate given in cubic feet per min. can be converted into cubic metres per min. by multiplying by 0.0283 M.P.L.

NY for uranium in air is 50 micrograms U^{238} per cubic metre of air.

1 R.E.M.(roentgen equivalent man) = 1 R.A.D. in tissue/R.B.E. R.B.E. is relative biological efficiency; R.A.D. is radiation when delivered to a given mass and produces the same energy conversion as one roentgen delivers to the same mass of air (8.38×10^{-6}J).

2.4 Maximum Permissible Levels/Standard Tolerances for Radioactive Isotopes

β – tolerances for *drinking water* — 1.2×10^{-14} curies per cc of water.
C^{14} tolerance *in air* — 7.5×10^{-15} curies per cc of air.
Sr^{90} (strontium 90) — 8.0×10^{-17} curies per cc of air.
Pu (plutonium) — 2.2×10^{-18} curies per cc of air.
U238 (uranium) — 4.0×10^{-17} curies per cc of air.

These tolerances allow for a high margin of safety and may be increased up to factors of 100 without any ill effects. In all cases, however, time factors would have to be considered and the tolerance related to a period of say three months.

M.P.L. for long-half-life alpha emitters.

Po^{210} (polonium 210) — 2.0×10^{-16} curies per cc of air.
Rn^{226} (radon 226) — 8.0×10^{-18} curies per cc of air.
Ca^{45} (calcium 45) — 3.0×10^{-13} curies per cc of air.
I^{131} (iodine 131) — 3.0×10^{-15} curies per cc of air.
F^{18} (fluorine 18) — 1.0×10^{-10} curies per cc of air.
Co^{60} (cobalt 60) — 1.0×10^{-12} curies per cc of air.

In considering these tolerances two important items are:

1 — The time factor.
2 — The efficiency of measuring instruments.

Most of the tolerances and M.P.L.s quoted were originally from American standards. Recently many of these figures have been decreased, in some cases up to a factor of ten.

2.5 Radioactive Half-Life of Some Common Elements (Isotopes)

Element	Radiation	Half-Life
Argon A^{35}	β	1.88 secs.
A^{39}	β	4.0 mins.
A^{41}	$\beta\gamma$	110 mins.
Gold Au^{198}	γ	2.7 days
Bromine Br^{82}	$\beta\gamma$	34 days
Br^{83}		140.0 mins.
Br^{84}	β	30.0 mins.
Carbon C^{11}	β	20.5 mins.
C^{14}		> 10 yrs.
Cobalt Co^{60}	γ	5.2 yrs.
Chlorine Cl^{133}	β	2.4 secs.
Cl^{134}	β	3.3 mins.
Cl^{136}	β	10 yrs.
Cl^{138}	$\beta\gamma$	37 mins.
Fluorine F^{59}	γ	44 days
Iodine I^{124}	$\beta\gamma$	4 days
I^{126}	$\beta\gamma$	13 days
I^{131}	β	8 days
Potassium K^{38}	$\beta\gamma$	7.7 mins.
K^{40}		1.42×10 yrs.
Krypton Kr^{78-90}		13 secs. to 4.0 hrs.
Sodium Na^{24}		14.8 hrs.
Na^{22}	e^{-8}	3 yrs.
Xenon $Xe^{124-127}$	e^{-}	34 days
Xe^{133}		7 days

Note: e^{-} β emission.

2.6 Comparative Radiation Measures

The amount of radiation from X-rays is given here:

Chest X-ray (full plate) — 0.05 roentgens (1973)
Skull X-ray — 1.3 roentgens (1973)
Abdomen X-ray — 1.0 roentgens (1973)
Pregnancy X-ray (lateral) — 5.0 roentgens (1973)
Gastrointestinal — 0.7 roentgens (1973)

An average X-ray apparatus at 5-feet distance gives approximately 20 roentgens (when working) per minute. Background radiation is of the order of 0.0004 roentgens. A comparative measurement, for physical effects, is known as the R.E.P. or roentgen equivalent physical.

The following table shows the relative effects of various radiations:

	Relative Biological Efficiency	R.E.P. for Total-Body Radiation
X- & γ rays	1	0.5
β rays (beta)	1	0.5
α rays (alpha)	20	0.025
Slow neutrons	10	0.1
Fast neutrons	5	0.05

Note: These are relative physical effects.

α particles (or rays) only travel fractional distances in the body, or they may be stopped by protective clothing. β rays can penetrate about one centimetre of tissue. γ rays & X-rays can penetrate several centimetres of tissue.

SECTION 3 (BIOLIGICAL SECTION)

3.1 Toxic Effects of Some Chemicals Which Induce Blood Changes

Anaemia

This may be produced by the following chemicals, and others:

Aniline; Arsenic; Sulphurous organic compounds (e.g. CS_2); Carbon Tetrachloride; DDT; Hydrazine (and compounds); Toluene; Trichlorethylene; Trinitrotoluene; Vanadium; Radium; and most radioactive compounds.

Leucopenia (deficiency of white cells); Antimony; Arsenic; Carbon Tetrachloride; DDT; Manganese; Vanadium; Toluene and nitro-compounds; Radium; Uranium; and most radioactive compounds.

Leucocytosis

(This is an increase in the leucocytes in the blood. Leukaemia is a more advanced, chronic stage.)

The list of chemicals that induce changes is not a comprehensive one. A number of drugs can also induce blood changes: e.g. sulphurous organic compounds.

The presence of toxic metals in the blood indicates absorption but symptoms are more likely to correlate with concentration of the substance. Abnormal blood effects may vary from stippling of red cells; aniso-cytosis (large and small cells); cell fragility; or other irregularities. Lead, mercury insecticides, fumigants, thorium, radium, halogen chlorides, nitrous fumes, and hydrazine and its compounds are all capable of producing some of the above effects if absorbed into, or ingested in, the body.

Toxic Substances in Urine

Abnormalities which occur in urine may be the result of toxic substances. Albumen may be increased as a result of kidney damage. The following can be detected in urine: CS_2, CCl_4, chlorobenzene, DDT, naphthol and other organic halogen compounds, oxalic acid. Also arsenic, uranium, vanadium and other metals.

3.2 Normal Human Blood Values

Blood urea — 20-30 mg per 100 cc of blood.
Blood nitrogen — 9-17 mg per 100 cc of blood.
Total protein — 6-8 g per 100 cc of blood.
Serum bilirubin — 0.4-1.5 mg per 100 cc of serum.
Oxygen capacity — 18-25 cc per 100 cc of blood.

Calcium — 8.5-11.5 mg per 100 cc of serum.
Chlorides — 570-620 mg per 100 cc of plasma.
Potassium — 16-22 mg per 100 cc of serum.
Sodium — 315-330 mg per 100 cc of serum.
Phosphorus — 3-4 mg per 100 cc of serum.
Fluorine — Less than 0.01 mg per litre of serum.

Red blood cells — 4.5-6 million per cubic millimetre.
Haemoglobin — 14-18 g per 100 cc of blood.
Reticulocytes — 0.1-0.5% (a high value indicates active regeneration).
ESR (sedimentation rate) — 9-20 mm per hour.
Haematocrit — 40-48% cell volume.
MCV (mean value) — 80-94 cubic microns.
MCH (mean haemoglobin) — 27-32 micrograms.

Red cell fragility — Hymolysis starts in solution of 0.42% NaCl, complete in 0.32% NaCl solution.

3.3 Normal Urine Values

Colour: straw.
No turbidity.
pH 4.8 to 7.5, goes acid on standing.
SG 1.001 to 1.003.
Glucose: none.
Albumen: none.

There may be a few casts, R.B.C.s, leucocytes, and epithelial cells.
(Respectively to 9000, R.B.C.s to 1.5 million, epith. cells to 10000.)

The presence of porphyrins may indicate metal poisons.

3.4 Biological Stains

Basic Fuchsin

1 g in 200 ml 50% aqueous alcohol.

Borax Carmine

4 g borax and 3 g carmine in 200 ml 50% aqueous alcohol.

Carbol Fuchsin

Stain for striated muscle, etc. and some bacteria.

Basic fuchsin 1 g
Phenol 5 g
Alcohol 10 ml
Water dist. 100 ml

Dissolve the basic fuchsin in alcohol. Dissolve phenol in water, mix, stand and decant or filter.

Stain 1 min., wash (water), decolourise slightly (alcohol). Wash again, dry and mount.

Delafield's Haematoxylin

400 ml saturated sol. of alum (56.5 g).
4 g haematoxylin in 25 ml alcohol.
Add haematoxylin solution drop by drop to ammonium alum.
Allow to stand, exposed to light.
Cellulose stains blue. Lignocellulose stains red.

Gram's Stain

0.5 (aq.) methyl violet add Lugol's iodine +
0.5% neutral red (in 1% acetic acid).
Stain half minute with methyl violet.
Pour off and add fresh Lugol's iodine.
Leave half minute and then wash with alcohol twice.
Then add neutral red, 1% acetic acid; stain 15 seconds.
Wash in aq. dist. and dry.

Gram + diphtheria, staphylococci, streptococci and, pneumonia stains blue.
Gram + B. coli group, typhus and gonococci, stains red.

Lugol's Iodine

1 part iodine, 2 parts KI, 100 parts aq. distilled.

Stains cellulose yellow, cell nuclei stain brown. (Suitable for staining amoeba, etc.)

Neutral Red

Safranine in 7.5% alcohol.

Gentian Violet

1 g in 25% alcohol, with 3 ml aniline added (can be as an alternative to methyl violet in Gram's stain).

Eosin

1% aq. or 70% alcohol stains cytoplasm pink.

Cotton Blue

1% in 100 ml of lacto-phenol.
Lacto-phenol is equal parts (w/v) lactic acid, phenol, glycerine, water.

Nigrosine

(Aniline black) — Stains cancer cells and some blood cells.
(From aniline hydrochlorate) — HeLa cells give black outline.

The Romanowsky Stains: Wright's, Jenner's, Giemsa's, and Leishman's can be obtained ready-mixed; they are used in 0.15% solution. They are suitable for pathological fluids, and blood which may contain protozoa.

If the blue colour is too intense, it can be reduced by washing in acetic acid (about 1/1000). If too mauve, rinse in caustic soda, diluted 1/7000.

Leishman's Stain

Ready-mixed 0.15 g in 100 ml of methanol (acetone-free).

Leishman's Stain from Ingredients

Methylene blue (alk.) — 0.5%.
Eosin in alcohol — 0.5%.

Technique for Blood Slides

Dry the film. Cover with stain for one minute. Then add a double amount of buffered distilled water and mix. After 7 minutes pour off, cover again with buffered distilled water for 2 minutes. Rinse and dry.

Sudan Black

Sudan black 0.3% in alcohol.
Buffer, 15% (w/v) phenol in 30% alcohol. Add this to 100 ml of distilled water containing 0.3% of disodium hydrogen phosphate. Add 40 ml of buffer to 60 ml of sudan black and allow to stand and filter.

Technique for Staining (Blood Slides)

Fix slide in formalin vapour.
Stain one hour. Wash with 70% alcohol. Wash in water.
Dry and counterstain if desired.

Monocytes scattered +ve. Granulocytes, dense + dense +ve/ Lymphocytes and mega-karyocytes -ve.

M.I.F. Solution

Merthiolate tincture (1/1000) 40 ml.
Formaldehyde (40% formalin) 5 ml.
Glycerine 1 ml.

Add 15 parts of above to 1 part of Lugol's iodine immediately before use. (Suitable for permanent amounts of protozoa.)

Staining Protozoa

Fix with Schaudin's fluid which consists of 100 parts alcohol, 200 parts of (aq.) saturated solution $HgCl_2$ for 5 mins and warm to 60°C. Wash the film in alcohol and apply Lugol's iodine for 2 minutes. Rinse in alcohol to remove iodine.

Iron Haematoxylin

(a) Haematoxylin 1 g. Alcohol 100 m.

(b) Liquor ferri. perchloride 30% 21 ml.
Concentrated hydrochloric acid 1 ml.
Water distilled 100 ml.
Stain for 2 minutes.

After staining wash in water. Pass through alcohol and clear with
xylol. Mount in balsam or similar compound. Preparations may be
counterstained with the following:

Fuchsin, acid (sat. sol.) 1-3 parts.
Picric acid (sat. sol.) 100 parts.
Stain for 30 seconds.

Ziehl-Neelsen's Stain

Basic fuchsin 1 g.
Alcohol 10 ml.
Phenol 5% 90 ml.

Technique for Staining

Stain for 3 minutes in above. Then wash in 25% (aq.) sulphuric
acid. The stain is best poured on to the slide hot. Counterstain with
methylene blue (sat. sol.) for 1 minute.
Methylene blue consists of 30 ml (sat. solution in alcohol).
Pot. hydroxide 1% (aq.) 1 ml.
Distilled water to 100 ml.

Wash in water and dry.

Acid-fast bacteria, such as TB, stain pink.

Fontana's Silver Impregnation Method (for spirochaetes)

(a) Acetic acid 1% 1 ml.
Formalin 20% 20 ml.
Distilled water to 100 ml.

(b) Tannic acid 5% in 1 % phenol (aq.) distilled.

(c) Silver nitrate 0.25% (aq.) distilled.

Technique for Staining

Add solution (a) to the wet smear, leave 1 minute and then wash.
Add (b) to the slide and heat gently until steam begins to rise. Leave 1 minute and wash.
Add (c) to the slide and heat as before. Leave for 1 minute and then wash. Allow to dry.

Spirochaetes are stained brown.

3.5 Culture Solutions

Knop's Water-Culture Solution (for Plants)

Calcium nitrate 1 g.
Magnesium sulphate 0.25 g.
Potassium hydrogen phosphate 0.25 g.
Ferric chloride trace.
Potassium nitrate 0.25 g.
Water to one litre.
pH should be 5. This solution will support all stages of plant life.

Alternative Water-Culture Solution

Pot. nitrate 1 g.
Pot. dihydrogen phosphate 0.5 g.
Calcium sulphate 0.5 g.
Mag. sulphate 0.5 g.
Sodium chloride 0.5 g.
2 drops of ferric chloride solution.
Water to one litre.

Pasteur's Solution for the Culture of Yeasts

Cane sugar 15 g.
Ammonium tartrate 1 g.
Pot. phosphate 0.2 g.
Calcium phosphate 0.02 g.
Mag. sulphate 0.02 g.
Water to 100 ml.

3.6 Agar Media

Czapek-Dox Agar

Sucrose 30 g.
Sodium nitrate 2 g.
Pot. phosphate (K_2HPO_3) 1 g.
Mag. sulphate 0.5 g.
Pot. chloride 0.5 g.

Ferrous sulphate 0.01 g.
Agar 15.0 g.
Water to one litre.

Sterilise by heating for 30 minutes at 100°C on 3 consecutive days
or autoclave. This is mainly for the growth of fungi.

Tryptone Soya Agar

Amino acids (glycine, leucine, etc.) 1.5% (aq.).
Soya bean meal 0.5% w/v.
Sodium Chloride 0.5%.
Agar 1.5%.
Sterilise. This is suitable for the growth of most bacteria.

Malt Agar

Malt extract 20 g.
Agar 15 g.
Water to one litre 15 g.
Autoclave at 120°C for 20 minutes to sterilise.

3.7 Comparative Size of Some Biological Objects

	Microns	Milli-Microns	Angstrom Units
	μ	mμ	Å
Erythrocyte			
(average red blood cell)	7.5	7500	75000
Serratia Marscesens			
(bacteria)	0.75	750	7500
Rickettsia	0.47	470	4700
Psittacosis Virus			
(from birds)	0.15	150	1500
Influenza Virus	0.085	85	850
Polio Virus	0.027	27	270
Protein Molecule	0.0025	2.5	25
Hydrogen Molecule	0.0001	0.1	1.0

SECTION 4

4.1 Atomic Weights of Elements

1969 Revision (based on carbon-12)

Actinium — Ac — 227*
Aluminium — Al — 26.9815
Americium — Am — 243*
Antimony — Sb — 121.7
Argon — Ar — 39.94
Arsenic — As — 74.2916
Astatine — At — 210*
Barium — Ba — 137.3
Berkelium — Bk — 247*
Beryllium — Be — 9.01218
Bismuth — Bi — 208.9806
Boron — B — 10.81
Bromine — Br — 79.904
Cadmium — Cd — 112.40
Caesium — Cs — 132.9055
Calcium — Ca — 40.08
Californium — Cf — 251*
Carbon — C — 12.011
Cerium — Ce — 140.12
Chlorine — Cl — 35.453
Chromium — Cr — 51.996
Cobalt — Co — 58.9332
Copper — Cu — 63.54
Curium — Cm — 247*
Dysprosium — Dy — 162.5
Einsteinium — Es — 254*
Erbium — Er — 167.2

Europium — Eu — 151.96
Fermium — Fm — 257*
Fluorine — F — 18.9984
Francium — Fr — 223*
Gadolinium — Gd — 157.2
Gallium — Ga — 69.72
Germanium — Ge — 72.5
Gold — Au — 196.9665
Hafnium — Hf — 178.4
Helium — He — 4.00260
Holmium — Ho — 164.9303
Hydrogen — H — 1.008
Indium — In — 114.82
Iodine — I — 126.9045
Iridium — Ir — 192.2
Iron — Fe — 55.84
Krypton — Kr — 83.80
Lanthanum — La — 138.905
Lawrencium — Lr — 256*
Lead — Pb — 207.2
Lithium — Li — 6.94
Lutetium — Lu — 174.97
Magnesium — Mg — 24.305
Manganese — Mn — 54.9380
Mendelevium — Md — 258*
Mercury — Hg — 200.5
Molybdenum — Mo — 95.9
Neodymium — Nd — 144.2
Neon — Ne — 20.17
Neptunium — Np — 237.0482
Nickel — Ni — 58.7
Niobium — Nb — 92.9064
Nitrogen — N — 14.0067
Nobelium — No — 255*

Osmium — Os — 190.2
Oxygen — O — 15.999
Palladium — Pd — 106.4
Phosphorus — P — 30.9738
Platinum — Pt — 195.0
Plutonium — Pu — 244*
Polonium — Po — 209*
Potassium — K — 39.10
Praseodymium — Pr — 140.9077
Promethium — Pm — 145*
Protactinium — Pa — 231.0359
Radium — Ra — 226.0254
Radon — Rn — 222*
Rhenium — Re — 186.2
Rhodium — Rh — 102.9055
Rubidium — Rb — 85.467
Ruthernium — Ru — 101.0
Samarium — Sm — 150.4
Scandium — Sc — 44.9559
Selenium — Se — 78.9
Silicon — Si — 28.08
Silver — Ag — 107.868
Sodium — Na — 22.9898
Strontium — Sr — 87.62
Sulphur — S — 32.06
Tantalum — Ta — 180.947
Technetium — Tc — 98.9062
Tellurium — Te — 127.6
Terbium — Tb — 158.925
Thallium — Tl — 204.3
Thorium — Th — 232.0381
Thulium — Tm — 168.9342
Tin — Sn — 118.6
Titanium — Ti — 47.9

Tungsten — W — 183.8
Uranium — U — 238.029
Vanadium — V — 50.941
Xenon — Xe — 131.30
Ytterbium — Yb — 173.0
Yttrium — Y — 88.9059
Zinc — Zn — 65.3
Zirconium — Zr — 91.22

* Designates the mass number of a selected isotope of the element. This may be the isotope with the longest half-life, or the predominant isotope of the element.

4.2 Density of Solutions

Density and Percentage by Weight for Sulphuric Acid

Density $D20°$	% by Weight H_2SO_4	Density $D20°$	% by Weight H_2SO_4	Density $D20°$	% by Weight H_2SO_4
1.000	0.2609	1.150	21.38	1.300	39.68
1.005	0.9855	1.155	22.03	1.305	40.25
1.010	1.731	1.160	22.76	1.310	40.82
1.015	2.485	1.165	23.31	1.315	41.39
1.020	3.242	1.170	23.95	1.320	41.95
1.025	4.000	1.175	24.58	1.325	42.51
1.030	4.746	1.180	25.21	1.330	43.07
1.035	5.493	1.185	25.84	1.335	43.62
1.040	6.237	1.190	26.47	1.340	44.17
1.045	6.956	1.195	27.10	1.345	44.72
1.050	7.704	1.200	27.72	1.350	45.26
1.055	8.415	1.205	28.33	1.355	45.80
1.060	9.129	1.210	28.95	1.360	46.33
1.065	9.843	1.215	29.57	1.365	46.86
1.070	10.56	1.220	30.18	1.370	47.39
1.075	11.26	1.225	30.79	1.375	47.92
1.080	11.96	1.230	31.40	1.380	48.45
1.085	12.66	1.235	32.01	1.385	48.97
1.090	13.36	1.240	32.61	1.390	49.48
1.095	14.04	1.245	33.22	1.395	49.99
1.100	14.73	1.250	33.82	1.400	50.50
1.105	15.41	1.255	34.42	1.405	51.01
1.110	16.08	1.260	35.01	1.410	51.52
1.115	16.76	1.265	35.60	1.415	52.02
1.120	17.43	1.270	36.19	1.420	52.51
1.125	18.09	1.275	36.78	1.425	53.01
1.130	18.76	1.280	37.36	1.430	53.50
1.135	19.42	1.285	37.95	1.435	54.00
1.140	20.08	1.290	38.53	1.440	54.49
1.145	20.73	1.295	39.10	1.445	54.97

Density D20°	% by Weight H_2SO_4	Density D20°	% by Weight H_2SO_4	Density D20°	% by Weight H_2SO_4
1.450	55.45	1.600	69.09	1.750	82.09
1.455	55.93	1.605	69.53	1.755	82.57
1.460	56.41	1.610	69.96	1.760	83.06
1.465	56.89	1.615	70.39	1.765	83.57
1.470	57.36	1.620	70.82	1.770	84.08
1.475	57.84	1.625	71.25	1.775	84.61
1.480	58.31	1.630	71.67	1.780	85.16
1.485	58.78	1.635	72.09	1.785	85.74
1.490	59.24	1.640	72.52	1.790	86.35
1.495	59.70	1.645	72.95	1.795	86.99
1.500	60.17	1.650	73.37	1.800	87.69
1.505	60.62	1.655	73.80	1.805	88.43
1.510	61.08	1.660	74.22	1.810	89.23
1.515	61.54	1.665	74.64	1.815	90.12
1.520	62.00	1.670	75.07	1.820	91.11
1.525	62.45	1.675	75.49	1.821	91.33
1.530	62.91	1.680	75.92	1.822	91.56
1.535	63.36	1.685	76.34	1.823	91.78
1.540	63.81	1.690	76.77	1.824	92.00
1.545	64.26	1.695	77.20	1.825	92.25
1.550	64.71	1.700	77.63	1.826	92.51
1.555	65.15	1.705	78.06	1.827	92.77
1.560	65.59	1.710	78.49	1.828	93.03
1.565	66.03	1.715	78.93	1.829	93.33
1.570	66.47	1.720	79.37	1.830	93.64
1.575	66.91	1.725	79.81	1.831	93.94
1.580	67.35	1.730	80.25	1.832	94.32
1.585	67.79	1.735	80.70	1.833	94.72
1.590	68.23	1.740	81.16		
1.595	68.66	1.745	81.62		

(Always carefully add acid to water)

Density and Percentage by Weight for Ammonia Solution

Density D20°	% by Weight NH_3	Density D20°	% by Weight NH_3	Density D20°	% by Weight NH_3
0.998	0.0465	0.958	9.87	0.918	21.50
0.996	0.512	0.956	10.405	0.916	22.125
0.994	0.977	0.954	10.95	0.914	22.75
0.992	1.43	0.952	11.49	0.912	23.39
0.990	1.89	0.950	12.08	0.910	24.03
0.988	2.35	0.948	12.58	0.908	24.68
0.986	2.82	0.946	13.14	0.906	25.33
0.984	3.30	0.944	13.71	0.904	26.00
0.982	3.78	0.942	14.29	0.902	26.67
0.980	4.27	0.940	14.88	0.900	27.33
0.978	4.76	0.938	15.47	0.898	28.00
0.976	5.25	0.936	16.06	0.896	28.67
0.974	5.75	0.934	16.65	0.894	29.33
0.972	6.25	0.932	17.24	0.892	30.00
0.970	6.75	0.930	17.85	0.890	30.685
0.968	7.26	0.928	18.45	0.888	31.37
0.966	7.77	0.926	19.06	0.886	32.09
0.964	8.29	0.924	19.67	0.884	32.84
0.962	8.82	0.922	20.27	0.882	33.595
0.960	9.34	0.920	20.88	0.880	34.35

Density and Percentage by Weight for Sodium Hydroxide

Density D20°	% by Weight NaOH	Density D20°	% by Weight NaOH	Density D20°	% by Weight NaOH
1.000	0.159	1.180	16.44	1.360	33.06
1.010	1.045	1.190	17.345	1.370	34.03
1.020	1.94	1.200	18.255	1.380	35.01
1.030	2.84	1.210	19.16	1.390	36.00
1.040	3.745	1.220	20.07	1.400	36.99
1.050	4.655	1.230	20.98	1.410	37.99
1.060	5.56	1.240	21.90	1.420	38.99
1.070	6.47	1.250	22.82	1.430	40.00
1.080	7.38	1.260	22.73	1.440	41.03
1.090	8.28	1.270	24.645	1.450	42.07
1.100	9.19	1.280	25.56	1.460	43.12
1.110	10.10	1.290	26.48	1.470	44.17
1.120	11.01	1.300	27.41	1.480	45.22
1.130	11.92	1.310	28.33	1.490	46.27
1.140	12.83	1.320	29.26	1.500	47.33
1.150	13.73	1.330	30.20	1.510	48.38
1.160	14.64	1.340	31.14	1.520	49.44
1.170	15.54	1.350	32.10	1.530	50.50

4.3 Solubility of Chemical Radicals in 100 g of Distilled Water at 20°C

Key:
i — insoluble   s — soluble   vs — very soluble
ss — slightly soluble   app — approximate value

	Acetate	Arsenate	Arsenite	Borate	Bromide
Al	i	i	i	i	vs
NH₄	150	s	vs	8	74
Sb		i	i		
Ba		0.055	i	i	104
Bi		i	i	i	
Cd	vs	i	i	i	102.3
Ca	34.7	i	i	0.2	0.5
Cr		i		i	200
Co	s	i	i	i	60 app
Cu.		i			i
Cu..	7.2		i	i	vs
Fe..	vs	i	i		112.1
Fe...	i	i	ss	i	
Pb	50	i	i	i	0.8
Li	280 app	i	i	vs	183
Mg	vs	i	i	i	96.5
Mn	3	i	i	i	141.8
Hg.	0.7 app	i	i		i
Hg..	12 app	i	i		1.0 app
Ni	16.6	i	i	i	133.2
K	240	20	s	70 app	64.4
Ag	1.0 app	i		0.9	i
Na	26	28	vs	2.8	79.5
Sr		0.284	ss	77	102.4
Zn	40 app	i	i	i	445

Key:
i — insoluble   s — soluble   vs — very soluble
ss — slightly soluble   app — approximate value

	Carbonate	Chlorate	Chloride	Chromate	Cyanide
Al	i	vs	70	i	i
NH$_4$	20 app	s	30 app	30 app	s
Sb			920		
Ba	0.0023	338	44.5	i	80
Bi	i			i	
Cd	i	399	167		1.7
Ca	0.0065	170 app	74.5	16 app	s
Cr			130		
Co		558			i
Cu.	i		i	i	i
Cu..		207			
Fe..	i		70.8		
Fe...			91.85		
Pb	i	151.3	1.0	i	ss
Li	1.50	301	80.7	130 app	
Mg	0.013	vs	167	211.5	
Mn	i		150		
Hg.	i	s	i	ss	
Hg..	i	25	5.7	ss	12.0 app
Ni	i	134.1	170 app		i
K	112	7	34.5	62.25	s
Ag	i	10	i	i	i
Na	21 app	99	35.97	89	s
Sr	i	174.9	52.9	0.12	ss
Zn	0.001	200.2	360		i

Key:
i — insoluble   s — soluble   vs — very soluble
ss — slightly soluble   app — approximate value

	Fluoride	Hydroxide	Iodine	Nitrate	Oxalate
Al	i	i	vs	60.1 app	i
NH$_4$	vs	53.0	170	192	4.2
Sb	25				
Ba	0.163	7.43	203.1	9.2	0.01
Bi			i		i
Cd	4.36		86.2	127	0.0034
Ca	0.0016	ss	205.7	129.1	i
Cr	vs	0.165		s	i
Co	ss	i	159 (ous) 31.91	133 (ous) 108.6	i
Cu.	i	i	i		i
Cu..	ss	i		123	i
Fe..	ss	i		300	0.02
Fe...	ss	0.14		vs	vs
Pb	0.064	i	0.065	52.3	i
Li	0.27	12.5	164	138	8
Mg	0.0076	i	140.8	200	0.07
Mn	i		vs	426.4	0.05
Hg.			0.042	vs	i
Hg..		i	i	vs	i
Ni	0.02	i	146.9	230 app	
K	92.3	112	144.2	33.02	33 app
Ag	182	i	i	215	i
Na	4.5	109	179.3	88.1	3.7
Sr	0.011	40 app	177.8	70.8	0.004
Zn	1.6	i	497.0	320.5 app	i

Key:
i — insoluble   s — soluble   vs — very soluble
ss — slightly soluble   app — approximate value

	Phosphate	Oxide	Sulphate	Tartrate
Al	i	i		
NH₄	25		75.4	s
Sb		i		
Ba	0.01	1.5	i	0.028
Bi	i	i		
Cd	i		76.6	
Ca	0.01		0.203	0.0185
Cr	i		120	
Co	i	i	60 (ous) 36.2	
Cu.	i	i	36	
Cu..	ss	i		0.042
Fe..	i	i	62	0.870 app
Fe...	i	i	vs	
Pb	i	i	0.004	0.011
Li	0.04	5.22	43.5	
Mg	i	i	76.9	0.8
Mn		i	105.3 app	
Hg.	ss	i	0.06	
Hg..	i	i	i	
Ni	i	i	75	
K			11.18	0.5
Ag	i	i	0.75	0.2
Na	7.8		50	29
Sr	i		0.01	0.2
Zn	i	i	161.5	0.022

4.4 Buffer Solutions Used in Electrophoresis and Other Experiments

Veronal Buffer

1.8 g/litre — diethyl barbituric acid.
10.3 g/litre — sodium diethyl barbiturate.
This gives a pH of 8.6.
Suitable for amino-acid separation by electrophoresis.

Phosphate Buffer

2.2 g/litre — anhydrous disodium phosphate.
0.6 g/litre — monosodium phosphate (monohydrate).
Mix in equal volumes to obtain a pH of 8.6.

Borate Buffer

8.8 g/litre — sodium borate.
4.65 g/litre — boric acid.
Mix in equal volumes to obtain a pH of 8.6.
Suitable for separation of sugars by electrophoresis.

Phthalate Buffer

5.10 g/litre — potassium bi-phthalate.
0.86 g/litre — sodium hydroxide.
Mix equal volumes to obtain a pH of 5.9.

"Buffer Solutions" (Clark and Lubs)

A: 50 cc M/5 boric acid + potassium chloride* + x cc N/5 NaOH diluted to 200 cc (*12.405 gm H_3BO_3 and 14.911 g potassium chloride per litre).

B: 50 cc M/5 KCl + x cc N/5 HCl diluted to 200 cc.

C: 50 cc M/5 potassium dihydrogen phosphate + x cc N/5 NaOH diluted to 200 cc.

A		B		C	
pH	x	pH	x	pH	x
7.8	2.61	1.2	64.5	5.8	3.72
8.0	3.97	1.4	41.5	6.0	5.70
8.2	5.90	1.6	26.3	6.2	8.60
8.4	8.50	1.8	16.6	6.4	12.60
8.6	12.00	2.0	10.6	6.6	17.80
8.8	16.30	2.2	6.7	6.8	23.65
9.0	21.30			7.0	29.63
9.2	26.70			7.2	35.00
9.4	32.00			7.4	39.50
9.6	36.85			7.6	42.80
9.8	40.80			7.8	45.20
10.0	43.90			8.0	46.80

4.5 Simple Tests for the Identification of Minerals

Useful in those parts of the world where there are limited facilities.

Observation

Colourless crystals — may be copper, iron, chromium, cobalt, nickel.
Yellow/green — may be iron salts, chromium.
Pale pink — may be manganese salts.
Rose pink — may be cobalt (blue if anhydrous).
Green/blue — may be copper.

Heated on a Charcoal Block with a Blowpipe

Make a small hollow on the block of charcoal, mix the compound with powdered charcoal, and heat with the blowpipe.

Observe the residue:

Blue mass — may be aluminium.
Bright green — may be zinc.
Dirty green — may be tin.
Pale pink — may be manganese.

Then Heated on the Charcoal Block and Mixed with a Small Amount of Anhydrous Sodium Carbonate

Observe the residue:

Metal spangles — may be copper.
Metal spangles with yellow incrustation — may be lead, tin.
Metal beads, no incrustation — may be silver, gold.
Metal, slight incrustation — may be tin (stannic).

Orange incrustation — may be bismuth.
Odour of garlic (poisonous) — may be arsenic.
White incrustation — may be antimony, zinc.

Simple Identification of Minerals
Using Platinum-Wire Flame Tests

Sometimes better results are obtained if a small amount of concentrated hydrochloric acid is mixed with the powdered mineral.

Colour of Flame	Indication
Green (apple)	Borate or barium
Crimson	Strontium
Electric blue	Copper or borate
Violet (lilac)	Potassium
Yellow (persistent)	Sodium
Orange (red)	Calcium
Livid blue	Arsenic, lead, antimony, etc.

Substances Mixed with Borax, and Heated Again
on a Platinum Loop

A thin glass rod may also be used, in lieu of platinum wire.

Reducing Flame	Oxidising Flame	Indication
Opaque	Blue/green	Copper
Bottle green	Yellow	Iron
Green	Yellow/green	Chromium
Violet	Colourless	Manganese
Blue	Colourless	Cobalt
Grey	Brown/violet	Nickel

Confirmatory tests may be performed, but usually a systemic procedure for qualitative analysis is necessary.

4.6 Mineralogical Liquids

Liquid	Density (g/cc)
Iodoethane (C_2H_5I)	1.94
Bromoform	2.9
1.1.2.2. Tetrabromoethane ($CHBR_2.CHBR_2$)	2.9
Mercuric potassium iodide solution ($HgI_2.2KI$ aq.)	3.1
Cadmium tungsto-borate saturated solution	3.2
Di-iodomethane	3.3
Barium mercuric iodide (BaI_2HgI_2 aq.)	3.5

Minerals	Density (g/cc)
Zircon	4 to 4.8
Carbon	3.5 +
Topaz	3.5 +
Diamond	3.5 +
Flint glass	3.5
Glass (soda)	2.5
Dolomite	2.8
Aluminium alloy	2.7
Quartz	2.6
Silicon	2.3
Boron	2.3
Opal	1.7 to 2.2
Magnesium	1.74

4.7 Standard Wire Gauge (S.W.G.) and Sheet-Metal Gauge (British)

S.W.G. No.	Diameter/Thickness (inches)	Diameter (mm)
1	0.300	7.620
5	0.212	5.385
10	0.128	3.251
12	0.104	2.641
14	0.080	2.032
16	0.064	1.626
18	0.048	1.219
20	0.036	0.914
22	0.028	0.711
24	0.022	0.559
26	0.018	0.457
28	0.048	0.376
30	0.0124	0.315
32	0.0108	0.274
34	0.0092	0.234
36	0.0076	0.1930
38	0.0060	0.1524
40	0.0048	0.1219
42	0.0040	0.1016

4.8 Compressed-Gas Cylinder Identification

British Colour Codes

Gas	Cylinder
Acetylene	Maroon
Air	Grey
Ammonia	Black, red or yellow band
Argon	Blue
Carbon dioxide	Black, tropical white or silver
Carbon dioxide (medical)	Grey
Carbon dioxide (for snow)	Green
Carbon monoxide	Red, yellow band
Chlorine	Yellow
Ethylene	Mauve, red band
'Freon', dichloro-difluoromethane (Formerly used in refrigs.)	Both ends mauve
Helium	Brown
Hydrogen	Red
Methane	Red
Nitrous oxide (medical)	Mauve
Oxygen (medical)	Black, white band
Oxygen & CO_2 mixture (medical)	Black, grey & white band
Sulphur dioxide	Green, yellow band

4.9 Useful Chemical Reactions and Equations

The first two or three of each series represents present or former commercial methods. Naturally occurring ores are also given.

Aluminium

$Al_2O_3.2H_2O$ (bauxite). $CaAl_2Si_3O_8$ (felspars). Na_3AlF_6 (cryolite).

$2Al_2O_3 + 2NaOH = 2NaAlO_2 + H_2O + Al(OH)_2$
$Al_2O_3 + 3C + N_2 = 3CO + 2AlN$ (electric method)
$2Al(OH)_3 \rightarrow Al_2O_3 + 3H_2O$ (1000°C or 4 atm150°C)

$AlN + 3H_2O = NH_3 + Al(OH)_3$
$3Na + AlCl_3 = Al + 3NaCl$
$2Al + 6HCl = 2AlCl_3 + 3H_2$
$2Al + 2NaOH + 2H_2O = 2NaAlO_2 + 2H_2$

$AlCl_3 + 3NH_4OH = Al(OH)_3 + 2NH_4Cl$
$Al_2O_3 + H_2O = 2HAlO_2$ (acid)
$H_3AlO_3 + 3NaOH = Na_3AlO_3 + 3H_2O$
$2Al(OH)_3 + 3H_2SO_4 = Al_2(SO_4)_3 + 6H_2O$

Silver

Ag_2S (argentite). $AgCl$ (horn silver). $3Ag_2S.Sb_2S_3$ (pyogyrite).

$Ag_2S + 2CuCl_2 = 2CuCl + 2AgCl + S$
$Ag_2S + 2CuCl = Cu_2S + 2AgCl$
$2AgCl + 2Hg = Hg_2Cl + 2Ag$

$AgCl + Hg = HgCl + Ag$
$Ag + HNO_3 + AgNO_3$ (Stass process)
$2AgNO_3 + 2NaOH = Ag_2O + 2NaNO_3 + H_2O$
$Ag_2O + NH_3 + Ag(NH_3)_2OH$ (fulminate)
$2KCN + AgCl = KAg(CN)_2$ soluble $+ KCl$
$Ag_2S + 2NaCl = 2AgCl + Na_2S$ (Mexican process)

Boron

$Ca_2B_6O_{11}$ (colemanite).

$Ca_2B_6O_{11} + 2Na_2CO_3 = Na_2B_4O_7 + 2CaCO_3 + 2NaBO_2$
$4NaBO_2 + CO_2 = Na_2B_4O_7 + Na_2CO_3$
$Na_2B_4O_7 + 7H_2O = 2NaOH + 4H_3BO_3$
$2B_2O_3 + 3Mg = Mg_3(BO_3)_2 + 2B$ extraction
$KBF_4 + 3K = 4KF + B$

$Ba(ClO)_2 + H_2SO_4 = BaSO_4 + 2HClO$
$Ba(ClO_3)_2 + H_2SO_4 = BaSO_4 + 2HClO_3$ (chloric acid).
$BCl_3 + 3K = 3KCl + B$
$2B + 2NaOH + 2H_2O = 2NaBO + 3H_2$

Calcium

$CaOSiO_2$ (hydraulic lime).
$CaAl_2Si_2O_8$ (limefelspar). $CaMg(SiO_3)_2$ (diopside).

$CaO + 3C + CaC_2 + C$
$3CaOCl_2 + 2NH_3 = 3CaCl_2 + 3H_2O + N_2$
$CaC_2 + N_2 \Leftrightarrow CaCN_2 + C$
$\qquad 800°C$

$CaC_2 + H_2O = Ca(OH)_2 + C_2H_2$
$6Ca(OH)_2 + 6Cl = 5CaCl_2 + Ca(ClO_3)_2 + 6H_2O$

$2Ca(OH)_2 + 2Cl_2 = Ca(ClO)_2 + 2H_2O + CaCl_2$

$6Ca(OH)_2$ dry $+ 3Cl_2 = CaClO_3 + CaCl.Ca(OH)_2$

$3Ca(OH)_2 + 2Cl_2 = Ca(OCl)_2 + CaCl_2.Ca(OH)_2.H_2O$

$Ca(HCO_3)_2$ heat $\rightarrow CaCO_3 + H_2O + CO_2$

$9CaCO_3 + Al_2O_3.2SiO_2.2H_2O$ heat $\rightarrow 3CaOAl_2O_3 + 6CaSiO_2 +$
  $9Ca + 2H_2O$

Copper

$CuFeS_2$ (copper pyrites). Cu_2S (glance).
Cu_2O (cuprite). $CuCO_3Cu(OH)_2$ (malachite).

$4CuFeS_2 + 9O_2 = 2Cu_2S + 6SO_2 + 2Fe_2O_3$ (reverb. furnace)

$3Fe_2O_3 + FeS = 7FeO + SO_2$

$2FeO + SiO_2 = FeSiO_4$

$CuFeS_2 + O_2 = CuSO_4 + FeSO_4$

$CuSO_4 + Fe$ (scrap) $= Cu + FeSO_4$

$CuO + Cu$ heat $\rightarrow Cu_2O$

$CuSO_4 + 2NaOH = Cu(OH)_2 + Na_2SO_4$

$Cu(OH)_2 + NH_4OH = Cu(NH_3)_4(OH)_2$ (Schweitzer's solution
  dissolves celulose)

$CuCl_2 + Cu + HCl$ conc. $\rightarrow CuCl$

$Cu + 4HNO_3 = Cu(NO_3)_2 + 2H_2O + 2NO_2$

$2CuSO_4 + 4KI = 2CuI + I_2 + 2K_2SO_4$

$CuSO_4 + 2KCN = Cu(CN)_2 + K_2SO_4$

$Cu + Cl_2 \Leftrightarrow CuCl_2$

$2CuSO_4 + 2Na_2CO_3 + H_2O \rightarrow CuOH_2.CuCO_3 + CO_2 + Na_2SO_4$

$2CuSO_4 + 4KI = Cu_2I_2 + 2K_2SO_4 + I_2$

Chromium

$PbCrO_4$ (crocoite). $Fe(CrO_3)_2$ (chromite).

$Fe(CrO_2)_2 + 4C = Fe + 2Cr + 4CO$ (electric)

$2Cr_2O_3 + O_2 \text{ heat} \rightarrow 4CrO_2 \rightarrow CrO$

$K_2Cr_2O_7 + H_2SO_4 = K_2SO_4 + 2CrO_3 + H_2O$

$K_2CrO_4 + H_2SO_4 = K_2SO_4 + CrO_3 + H_2O$

$CrO_3 + H_2O \rightarrow H_2CrO_4$ (chromic acid)

$Cr_2(SO_4)_3 + 6NH_4OH = 2Cr(OH)_3 + 3(NH_4)2SO_4$

$3NaOH + CrCl_3 = Cr(OH)_3 + 3NaCl$

$Cr(OH)_3 + \text{heat} \rightarrow Cr_2O_3$

$Cr(OH)_3 + 3HCl = CrCl_3 + 3H_2O$

$Cr(OH)_3 \text{ in } H_2SO_4 \rightarrow Cr_2(SO_4)_3$ violet

$H_2CrO_4 + SO_2 \rightarrow Cr_2(SO_4)_3$ green

$4Fe(CrO_2)_2 + 8K_2CO_3 + 7O_2 = 8K_2CrO_4 + 2Fe_2O_3 + 8CO_2$

$2K_2CrO_4 + H_2SO_4 = K_2SO_4 + K_2Cr_2O_7 + H_2O$

$K_2Cr_2O_7 + 2KOH = 2K_2CrO_4 + H_2O$

$K_2Cr_2O_7 + 2NH_4Cl \Leftrightarrow (NH_4)_2Cr_2O_7 + 2KCl$

$(NH_4)_2Cr_2O_7 \text{ heat} \Leftrightarrow N_2 + H_2O + Cr_2O_3$

Hydrogen (Peroxide) & Water

$H_2O_2 + 2HI = 2H_2O + I_2$

$H_2O_2 + O_3 = H_2O + 2O_2$

$3H_2 + N_2 \Leftrightarrow 2NH_3 + \text{heat}$

$H_2O + C = CO + H_2$ (water gas)

$H_2O + CO = CO_2 + H_2$

Iodine

$2NaI + MnO_2 + 3H_2SO_4 = 2NaHSO_4 + MnSO_4 + H_2O + I_2$

$2NaIO_3 + 5NaHSO_3 = 3NaHSO_4 + 2Na_2SO_4 + H_2O + I_2$

$2KI + 3H_2SO_4 + MnO_2 = 2KHSO_4 + MnSO_4 + H_2O + I_2$

$2KI + 2HNO_3 = 2KOH + I_2 + 2NO$

$HI + 2HNO_3 = 2KOH + I_2 + 2NO$

$I_2 + SO_2 + 2H_2O = 2HI + H_2SO_4$

Iron

Fe_2O_3 (haematite). Fe_3O_4 (magnetite). FeS_2 iron pyrites.

$Fe_2O_3 + 3CO = 2Fe + 3CO_2$
$4FeS_2 + 12O_2 = 2Fe_2O_3 + 8SO_2$
$FeSO_4 + 2KCN = Fe(CN)_2 + K_2SO_4$
$Fe(CN)_2 + 4KCN = K_4Fe(CN)_6$
$FeSO_4 + 6KCN = K_4Fe(CN)_6, + K_2SO_4$
$3Fe + SO_2 = 2FeO + FeS$
$FeSO_4$ (-ous) heat $\rightarrow Fe_2O_3 + SO_2 + SO_3$
$Fe_2(SO_4)_3$ heat $\rightarrow Fe_2O_3 + 3SO_3$
$2Fe_2O_3 + 3Si = 3SiO_2 + 4Fe$
$2Fe_2O_3 + 3S = 2SO_2 + 4Fe$ Siemens
$Fe_2O_3 + 3Mn = 3MnO + 2Fe$
$FeO + Fe_3C = 4Fe + CO$

$Fe(OH)_2$ heat $\rightarrow Fe_2O_3$
$Fe + 2HCl$ (dil.) $\rightarrow FeCl_2 + H_2$
$Fe + H_2SO_4$ (dil.) $\rightarrow FeSO_4 + H_2$
Fe oxalate heat $\rightarrow FeO$
$2FeS_2 + 2H_2O + 7O_2 \rightarrow 2FeSO_4 + 2H_2SO_4$
$2Fe + 4(OH) \rightarrow 2Fe(OH)_2 + O_2 + Fe_2O_3\ H_2O$ rusting
$4Fe + 10HNO_3 = 4Fe(NO_3)_2 + NH_4NO_3 + 3H_2O$
$FeSO_4 + 2NaOH \rightarrow Na_2SO_4 + Fe(OH)_2$
$2FeSO_4$ heat $\rightarrow Fe_2O_3 + SO_2 + SO_3$
$6FeSO_4 + 3H_2SO_4 + 2HNO_3 \rightarrow 3Fe_2(SO_4)_3 + 4H_2O + 2NO$
    brown ring test

$2FeCl_2 + Cl_2 = 2FeCl_3$
$2FeCl_3 + H_2 = 2FeCl_2 + 2HCl$
$FeCl_3 + (NH_4)_2S$ xs alk. $= Fe_2S_3$
$3Fe + SO_2 = 2FeO + Fe$

$FeSO_4 + 2KCN = Fe(CN)_2 + K_2SO_4$
$FeSO_4 + 6KCN = K_4Fe(CN)_6 + K_2SO_4$
$Fe(CN)_2 + 4KCN = K_4Fe(CN)_6$
$2FeSO_4 + H_2O_2 + H_2SO_4 = Fe_2(SO_4)_2 + 3H_2O$

Lead

$PbSiF_6 + H_2SiF_6$ (electric)
$2PbO_2$ heat $\rightarrow PbO$
$PbO + 2KOH \rightarrow K_2PbO_2$ (plumbite) $+ H_2O$
$PbO + 2HNO_3 \rightarrow Pb(NO_3)_2 + H_2O$
$PbO + O_2 \rightarrow 2Pb_3O_4$
$\qquad$ 400°C

$2Pb_3O_4$ heat $\rightarrow 6PbO + O_2$
$PbO_2.2PbO + 4HNO_3 = PbO_2 + 2Pb(NO_3)_2 + 2H_2O$
$PbO_2 + HCl$ (conc.) $= PbCl_4 + 2H_2O$
$PbO + O_2 + NaOH + NaOCl \rightarrow Pb_3O_4$
$Pb(NO_3)_2 + 2HCl = PbCl_3 + 2HNO_3$
$PbS + 2PbO = 3Pb + SO_2$
$PbS + PbSO_4 = 2Pb + 2SO_2$

Manganese

MN_3O (housemanite). Mn_2O_3 (braunite). MnO_2 (pyrolusite).

$Mn_2O_3 + 2C = Al_2O_3 + 2Mn$ (thermite)
$MnO_2 + 2C = Mn + 2CO$ (electric)
$MnO_2 + 2CO = Mn + 2CO_2$ (blast)
$MnSO_4 + 2NaOH = Mn(OH)_2 + Na_2SO_4$
$2MnO_2 + 2H_2SO_4 = 2MnSO_4 + 2H_2O + O_2$
$MnCO_3$ heat $\rightarrow MnO + CO_2$
$Mn(NO_3)_2$ heat $\rightarrow MnO_2 + 2NO_2$

$2MnO + O_2 = 2MnO_2$
$MnCO_3 + HNO_3 = Mn(NO_2)_2$
$MnSO_4 + (NH_4)_2S = MnS + (NH_4)_2SO_4$
$6MnO$ air, heat $\rightarrow 2MnSO_4$
$2MnO_2 + 4KOH + O_2 = 2K_2MnO_4 + 2H_2O$
$3K_2MnO_4 + 2CO_2 = 2KMnO_4 + 2K_2CO_3 + MnO_2$
$2K_2MnO_4 + Cl_2 = 2KMnO_4 + 2KCl$

Phosphorous

$Ca_3(PO_4)_2$ (calcium phosphate). $CaF_2SCa_3(PO_4)_2$ (apatite).

$2Ca_3(PO_4)_2 + 6SiO_2 + 10C = 6CaSiO_2 + 10CO + P_4$
$Ca_3(PO_4)_2 + 3SiO_2 = 3CaSiO_3 + P_2O_5$
$P_2O_5 + 5C = 2P + 5CO$
$2HNO_3 + P_2O_5 = 2HPO_3 + N_2O_5$
$2H_3PO_4$ 250°C $\rightarrow H_4P_2O_7$ (pyro acid) + H_2O
$2Na_2HPO_4 \rightarrow Na_4P_2O_7 + H_2O$
$4P + 3CuSO_4 + 6H_2O = Cu_3P_2 + 2H_3PO_3 + 3H_2SO_4$
$Cu_3P_2 + 5CuSO_4 + 8H_2O = 8Cu + 5H_2SO_4 + 2H_3PO_4$
P_2O_5 + cold $H_2O \rightarrow 2HPO_3$ meta
P_2O_5 + hot $3H_2O \rightarrow 2H_3PO_4$ ortho
$H_2SO_4 + P_2O_5 = 2HPO_3 + SO_3$
$2HNO_3 + P_2O_5 = 2HPO_3 + N_2O_5$
$3Ba(OH)_2 + 8P + 6H_2O = 2PH_3 + 3Ba(H_2PO_2)_2$
$Ba(H_2PO_2)_2 + H_2SO_4 = BaSO_4 + 2H_3PO_2$
$H_3PO_2 + 2H_2O + CuSO_4 = H_3PO_4 + H_2SO_4 + CuH_2$
$P_2O_3 + 2H_2O = 2H_3PO_3$ (phosphorous acid)
$PCl_3 + 3H_2O = H_3PO_3 + 3HCl$
$2H_2PO_3 + H_2O = H_3PO_3 + H_3PO_4$
$P_2O_5 + 3H_2O = 3H_3PO_4$ (boil, converts meta to para)
$P + 5HNO_3 = H_3PO_4 + 5NO_2 + H_2O$
$H_3PO_4 \rightarrow HPO_3 + H_2O$
$2P + 3Cl_2 = 2PCl_3$

$PCl_3 + 3H_2O = H_3PO_3 + 3HCl$ (-ous)

$PCl_3 + 3C_2H_5OH \rightarrow H_3PO_3 + 2POCl_3 + 2HCl$

$H_2SO_4 + 2PCl_5 = SO_2Cl_2 + 2POCl_3 + HCl$

PBr_3 from Br — P (dist. from benzene)

$KClO_3 + 3PCl_3 = KCl + 3POCl_3$

$PCl_5 + H_2O = POCl_3 + 2HCl$

$POCl_3 + 3H_2O = H_3PO_4 + 3HCl$

$2P$ (white) $+ 3Cl_2 = 2PCl_3$

$P_2O_5 + 2HNO_3 = 2HPO_3 + N_2O_5$

Mercury

HgS (cinnabar). $HgCl$ (horn quicksilver).

$HgS + O_2 = Hg + SO_2$

$2HgO + 2Cl_2$ cool $\rightarrow HgO.HgCl_2 + Cl_2O$ (explosive monoxide).

$HgCl_2 + Hg = Hg_2Cl_2$ ($2HgCl$)

Hg_2Cl_2 distil $\rightarrow HgCl_2 + Hg$

$2HgNO_3 + 2HCl = Hg_2Cl_2 + 2HNO_3$

$Hg + I_2$ excess $= HgI_2$

$2Hg + 2H_2SO_4$ heat $\rightarrow HgSO_4 + SO_2 + 2H_2O$

$HgSO_4 + 2NaCl = HgCl_2 + Na_2SO_4$

Nickel

$NiAsS$ (nickel glance). $NiFe$ (pentlandite).

$CuFeS_2$ copper pyrites $+ 2\%$ Ni

$2NiS + 3O_2 = 2NiO + 2SO_2$

$Ni +$ water gas $350°C \rightarrow Ni(CO)_4$

$NiCl_2 + NaOCl + 2NaOH = NiO_2.H_2O$ (dioxide) $+ 3NaCl$

Potassium

$KAlSi_3O_8$ (potash felspar). $KClMgCl_2.6H_2O$ (carnallite).

$4KCl.MgCl_2 + 6H_2O = 3KCl + KCl.MgCl_2.6H_2O + 3MgCl_2$

$KCl + 3H_2O = KClO_3 + 3H_2$ electrolysis

$3MgCO_3 + 2KCl + CO_2 + 9H_2O = 2KHCO_3 + MgCO_3.4H_2O + K_2CO_3 + CO_2 + 5H_2O$

$2K + 2H_2O = 2KOH + H_2$ (explosive)

$2K + O_2 = K_2O_4$

$K_2O_4 + 2H_2O = 2KOH + H_2O + O_2$

$6Ca(OH)_2 + 6Cl_2 = 5CaCl + Ca(ClO_3)_2 + 6H_2O$

$Ca(ClO_3)_2 + 2KCl = CaCl_2 + 2KClO_2$

$2K + 2H_2O \rightarrow 2KOH + H_2$

K_2HgI_4 (Nessler's reagent).

$2KOH + 3Cl_2 = KClO_2 + KClO_3 + H_2O$

$2KOH + 3Cl_2 = 5KCl + KClO_3 + 3H_2O$

$6KOH + 3Cl_2 = 5KCl + KClO_3 + 3H_2O$

$4Fe(CrO_2)_2 + 8K_2CO_3 + 7O_2 = 8K_2CrO_4 + 2Fe_2O_3 + 8CO_2$

$Ca(ClO_3)_2 + 2KCl = CaCl_2 + 2KClO_3$

$2KCrO_2O_7 + H_2SO_4 = K_2SO_4 + K_2Cr_2O_7 + H_2O$

$K_2Cr_2O_7 + 2NH_4Cl \Leftrightarrow (NH_4)_2Cr_2O_7 + 2KCl$

$K_2Cr_2O_7 + 2KOH = 2K_2CrO_4 + H_2O$

$(NH_4)_2Cr_2O_7$ heat $\rightarrow N_2 + 4H_2O + Cr_2O_3$

$3K_2MnO_4 + 2CO_2 = 2KMnO_4 + 2K_2CO_3 + MnO_2$

$2K_2MnO_4 + Cl_2 = 2KMnO_4 + 2KCl$

$2K_2MnO_4 + O + H_2O \rightarrow 2KMnO_4 + 2KOH$ commercial (electrical)

$K_2MnO_4 + H_2O = MnO_2 + 2KOH + O_2$

$K_2Mn_2O_8 + 16HCl = 2KCl + 2MnCl_2 + 5Cl_2 + 8H_2O$

$KOI + H_2O_2 = KI + O_2 + H_2O$

$4K + 3SO_2 = K_2S_2O_3 + K_2SO_3$

$2KI + H_2O_2 = 2KOH + I_2$ (in presence of $FeSO_4 + O_2$)

Perchlorates

$KClO_3 + H_2SO_4$ cool $\rightarrow KHSO_4 + HClO_3$ (chloric acid)
$3HClO_3$ heat $\rightarrow HClO_4$ (perchloric acid) $+ H_2O + ClO_2$ explosive.
$2HClO_4P_2O_5 \rightarrow H_2O + Cl_2O_7$
$KClO_3 + H_2SO_4$ cool $\rightarrow KHSO_4 + HClO_3$ care

Sodium

$AlF_3.3NaF$ (cryolite). $NaAlSi_3O_8$ (soda felspar).

$2NaCl \rightarrow 2Na + Cl_2$ (electric method)
$3Na_2O_2 + 2C = 2Na_2CO_3 + 2Na$
$NaNO_3$ heat $\rightarrow NaNO_2 + O_2$

$NH_4OH + CO_2 = NH_4HCO_3$
$NH_4CO_3 + NaCl = NH_4Cl + NaHCO_3$

$NaCl + H_2SO_4$ heat $\rightarrow NaHSO_4 + HCl$
$NaHSO_4 + NaCl$ heat $\rightarrow Na_2SO_4 + HCl$
$Na_2SO_4 + 2C = Na_2S + 2CO_2$
$Na_2S + CaCO_3 = Na_2CO_3 + CaS$

$2Na + 2C_2H_5OH = 2C_2H_5O.Na + H_2$
$2C_2H_5O.Na + 2H_2O$ heat $\rightarrow 2C_2H_5OH + 2NaOH$
$Ba(OH)_2 + Na_2CO_4 + 2NaOH + BaCO_3$
$Na_2CO_3 + Ca(OH)_2 = CaCO_3 + 2NaOH$
$Na_2CO_3 + Fe_2O_3$ fused $\rightarrow Na_2Fe_2O_4 + CO_2$
$Na_2Fe_2O_4 + H_2O \rightarrow 2NaOH + 2Fe(OH)_3$
$Zn + 2NaOH = Na_2ZnO_2$ (zincate) $+ H_2$
$2B + 2NaOH + 2H_2O = 2NaBO_2 + 3H_2$
$2Al + 2NaOH + 2H_2O = 2NaAlO_3$ (aluminate) $+ 2H_2$
$Si + 2NaOH + H_2O = Na_2SiO_3 + 2H_2$
$Cl_2 + 2NaOH = NaCl + NaOCl + H_2O$

$4P + 3NaOH + 3H_2O = 3NaH_2PO_2$ (hypophosphite) PH_3
   Sp. inflammable gas

$2NaOH + NO + NO_2 = 2NaNO_2 + H_2O$

$3Cl_2 + 6NaOH = NaClO_3 + 5NaCl + 3H_2O$

$Na_2O_2 + CO = Na_2CO_3$

$3NaN_3 + NaNO_2 = 2Na_2O + 5N_2$

$2NaOH + MnSO_4 = Mn(OH)_2 + Na_2SO_4$

$2NaHSO_3$ heat $\rightarrow NaS_2O_3$ (metabisulphite) $+H_2O$

$Na_2SO_3 + S = Na_2S_2O_3$

$Na_2S_2O_3 + HCl = 2NaCl + S + SO_2 + H_2O$

$2NaHSO_3 + SO_3 + Zn = Na_2S_2O_4 + ZnSO_4 + H_2O$

$2Na_2S_2O_3 + I_2 = 2NaI + Na_2S_4O_6$ (tetrathionate)

$2NO_2 + 2NaOH = NaNO_2 + NaNO_3 + H_2O$

$NaOCl + NH_3 = NaOH + NH_2Cl$ (chloramine)

$NaNH_2 + N_2O = H_2O + NaN_3$

$2NH_3 + 2Na$ heat $\rightarrow 2NaNH_2 + H_2$

$NaOCl + 2KI + 2HCl = NaCl + 2KCl + H_2O + I_2$ volumetric

$Na_2S_2O_3 + AgCl = NaAgS_2O_3 + NaCl$

$Na_2CO_3 + Fe_2O_3$ fused $\rightarrow Na_2Fe_2O_4$ (ferrite) $+ CO_2$

$Na_2Fe_2O_4 + H_2O = 2NaOH + 2Fe(OH)_2$

Potassuim Chlorates

$2HClO_3 \rightarrow HClO_4 + 2ClO_2 + H_2O$

$6KOH + 3Cl_2 = 5KCl + KClO_3 + 3H_2O$

$4KClO_3$ heat $\rightarrow 3KClO_4 + KCl$

$KClO_4 + H_2SO_4$ reduced pressure $\rightarrow KHSO_4 + HClO_4$

$KClO_3 + 3PCl_3 = +3POCl_3 + KCl$

$10NO + 6KMnO_4 + 12H_2SO_4 = 10HNO_3 + 6KHSO_4 + 6MnSO_4 +$
   $4H_2O$

$KMnO_4 + 8HCl + 5K_4Fe(CN)_6 = 6KCl + MnCl_2 + 5K_3Fe(CN)_6 +$
   $4H_2O$

$K_4Fe(CN)_6 + 8H_2SO_4 + 6H_2O = 3(NH)_2SO_4 + 4KHSO_4 + FeSO_4 + H_2O$

dil. gives HCN

$FeSO_4 + 6KCN = K_4Fe(CN)_6 + K_2SO_4$

$4K + 3SO_2 = K_2S_2O_3 + K_2SO_3$

$2KI + H_2O + O_3 = 2KOH + O_2 + I_2$

$BCl_3 + 3K = 3KCl + B$

$KBF_4 + 3K = 4KF + B$

Sulphur

$H_2S + H_2O_2 = 2H_2O + S$

$6NaOH + 12S = Na_2S_2O_3 + 2Na_2S_5 + 3H_2O$

$Na_2S_2O_3 + 2HCl = 2NaCl + SO_2 + H_2O + S$ colloid

S_2 (molten) $+ 3Cl_2 = S_2Cl_2 + Cl_2$ (ice cold) $+ SCl_4$

$SO_2 + 2H_2S = 2S + 2H_2O$

$2S + SO_3$ (liq.) $= S_2O_3$

$S_2O_3 + H_2O = H_2S_2O_4$ (hydrosulphurous acid)

$2SO_2 + O_2$ cat. $\rightarrow 2SO_3$

$2S_2Cl_2 + 3H_2O = 2Cl_2 + H_2SO_4 + 3S$

$SO_3 + S_2Cl_2 = SOCl_2 + SO_2 + S$

$SO_2 + Cl_2$ camphor $\rightarrow SO_2Cl_2$

$2HOSO_2 + 2NO_2 + H_2O = 2H_2SO_4 + NO + NO_2$

$NO_2 + SO_2 + O_2 + H_2O =$ lead chamber acid

$2H_2SO_4 + NO + NO_2 = 2HOSO_2$ nitro-sulphuric $+ 2NO_2$ Glover towers $+ H_2O$

$SO_2Cl_2 + 2H_2O = H_2SO_4 + 2HCl$

$PCl_5 + SO_2 = SOCl_2 + POCl_3$

$H_2SO_4 + PCl_5 = SO_2Cl_2 + POCl_3 + H_2O$

$CS_2 + 3Cl_2$ cat. $= CCl_4 + S_2Cl_2$

$S + 6HNO_3$ boil $= H_2SO_4 + 6NO_2 + 2H_2O$

$4O_3 + H_2S = H_2SO_4 + 4O_2$

$H_2SO_4 + SO_3 = H_2S_2O_7$ (pyrosulphuric acid)

$H_2S + NO_2 = NO + H_2O + S$

$2Na_2HPO_4$ heat $\rightarrow Na_4P_2O_7 + H_2O$

$SO_2 + 2HNO_2 = H_2SO + 2NO$

Tin

$Sn + HNO_3$ (conc.) heat $\rightarrow SnO_2H_2O$ β (stannic acid)

$Sn + 2HCl$ (conc.) heat $\rightarrow SnCl_2 + H_2$

$Sn + 2H_2SO_4$ (conc.) heat $\rightarrow SnSO_4 + 2H_2O + SO_2$

$SnO_2 + 2C \rightarrow 2CO + Sn$

$4Sn + 10HNO_3$ (cold dil.) $= 4Sn(NO_3)_2 + NH_4NO_3 + 3H_2O$

$Sn + 4HNO_3 \rightarrow SnO_2.H_2O + 4NO_2 + H_2O$

$SnCl_2 + HNO_3 \rightarrow$ Bettendorf reagent

SnC_2O_4 heat $\rightarrow SnO + CO + CO_2$

$Sn + 2NaOH$ heat $= Na_2SnO_2$ (soluble) $+ 2H_2$

$SnO_2 + 2NaOH$ heat $= Na_2SnO_3$ (stannate) $+ H_2O$

$Sn(OH)_2 + 2NaOH = Na_2SnO_2 + 2H_2O$

$SnCl_2 + 2HgCl = Hg_2Cl_2 + SnCl_4$

$SnCl_2 + Hg_2Cl_2 = 2Hg + SnCl_4$

$3Sn + 7HNO_3$ (dil.) $= 3Sn(NO_3)_2 + NH_3OH + 2H_2O$

Zinc

ZnS (zinc blende). $ZnCO_3.ZnO$. $ZnO.Fe_2O_3$ (franklinite). $H_2Zn_2SiO_5$ (electric calamine).

$Zn + H_2SO_4 = ZnSO_4 + H_2$

$4Zn + 5H_2SO_4 = 4ZnSO_4 + H_2S + 4H_2O$

$3Zn + 4H_2SO_4 = 3ZnSO_4 + S + 4H_2O$

$Zn + 2H_2SO_4 = ZnSO_4 + SO_2 + H_2O$

$Zn + 2SO_2$ (sol. pt.) $\rightarrow ZnS_2O_4$ (hydrosulphite)

$Zn + 2HClO_4 = Zn(ClO_4)_2 + H_2$

$Zn + 4HNO_3 = Zn(NO_3)_2 + 2H_2O + 2NO_2$

$3Zn + 8HNO_3 = 3Zn(NO_3)_2 + 4H_2O + 2NO_2$

$4Zn + 10HNO_3 = 4Zn(NO_3)_2 + NH_4NO_3 + 2H_2O$

$4Zn + 9HNO_3$ (5%) $= 4Zn(NO_3)_2 + NH_3 + 3H_2O$

4.10 Chemical Names for Common Substances

Alum — Potassium aluminium sulphate

Alumina — Alumunium oxide Al_2O_3

Antichlor — Sodium thiosulphate $Na_2S_2O_3.5H_2O$

Antifibrin — Acetanilide

Antimony bloom — Antimony trioxide Sb_2O_3

Antimony black — Antimony sulphide Sb_2S_3

Aqua regia — Concentrated nitric acid and hydrochloric acid HNO_3 + 3HCl

Aspirin — Acetylsalicylic acid $C_6H_4(OCOCH)_3COOH$

Baking powder — Sodium bicarbonate $NaHCO_3$

Barium white — Barium sulphate $BaSO_4$

Bryta — Barium oxide BaO

Barytes — Barium sulphate $BaSO_4$

Bauxite — Hydrated alumina $Al_2O_3.2H_2O$

Bentonite — Aluminium sulphate (impure)

Benzol — Benzene (impure)

Black ash — Impure sodium carbonate

Bleaching powder — Calcium hypochlorite $CaOCl_2$

Blue vitriol (blue john) — Copper sulphate $CuSO_4.5H_2O$

Bone ash — Animal charcoal

Boracic acid — Boric acid H_3BO_3

Borax — Sodium tetraborate $Na_2B_4O_7.10H_2O$

Brimstone — Sulphur

Burnt lime — Calcium oxide

Calomel — Mercurous chloride Hg_2Cl_2

China clay — Aluminium Silicate

Chloramine T — Sodium toluene sulpho-chloramide

Chrome alum — Potassium chromium sulphate

Corn sugar — Glucose

Cream of tartar — Potassium hydrogen tartrate

Eau de Javelle — Potassium hypochlorite solution

Formalin — Formaldehyde 40% aq. solution

Lunar caustic — Silver nitrate (stick) $AgNO_3$

Marsh gas — Methane
Microcosmic salt — Sodium ammonium hydrogen phosphate
Milk of lime — Calcium hydroxide $Ca(OH)_2$
Milk of Magnesia — Magnesium hydroxide $Mg(OH)_2$
Mohr's salt — Ferrous ammonium sulphate
Muriatic acid — Hydrochloric acid

Naphtha — Petroleum distillate
Nordhausen acid — Fuming sulphuric acid

Oil of wintergreen — Methyl salicylate
Osmic acid — Osmium tetroxide

Paris green — Copper aceto-arsenite
Plaster of Paris — Calcium sulphate $CaSO_4.\frac{1}{2}H_2O$
Plumbago — Graphite

Quicklime — Calcium oxide CaO

Rectified spirit — Alcohol (ethyl) 90-95%
Rochelle salt — Potassium sodium tartrate

Sal ammoniac — Ammonium chloride NH_4Cl
Saltpetre — Potassium nitrate
Scheele's green — Copper hydrogen arsenite
Solvent naphtha — Coal tar distillate

Spirit of hartshorn -- Ammonia solution
Spirits of salt — Hydrochloric acid
Sugar of lead — Lead acetate
Sulphuric ether — Diethyl ether
Superphosphate — Calcium hydrogen phosphate (impure)

Talc — Hydrated magnesium silicate
Tartar emetic — Potassium antimony tartrate

Verdigris — Basic copper carbonate
Vitriol — Sulphuric acid (concentrated)

Washing soda — Sodium carbonate
Whiting — Calcium carbonate
Wood alcohol — Methyl alcohol

4.11 Some Useful Formulae (Cements and Waxes)

Glycerine Litharge Cement

Glycerine 6 parts.
Water 1 part.
Litharge, sufficient to make into a paste.

High-Melting-Point 'Wax'

Iron sulphide 1 part.
Zinc sulphide 1 part.
Lead sulphide 1 part.
Sulphur 1 part.
Melt and stir together then cool.
Melting point about 155°C.
Resistant to most acids and alkalis.

Glass Marking Ink

Water glass ($NaSiO_3$) 10 parts.
Barium sulphate 10 parts (or drawing ink for black marking).

Low-Melting Alloy

Lead 35 g.
Bismuth 65 g m.p. 92°C.
Lead 27 g, zinc 13 g, bismuth 48 g, cadmium 12 g.
Gives an alloy m.p. 69°C.

4.12 Hardness Scale, Battery Acid and Freezing Mixtures

Mohr's Scale of Hardness

Diamond	10
Corundum	9
Topaz	8
Quartz	7
Felspar	6
Calcite	3
Rock salt	2
Talc	1

Steel file 6.5: Window glass 5.5: Fingernail 2.3.

Battery Acid

Acid suitable for acid/lead accumulators can be prepared as follows:

481 ml of concentrated sulphuric acid is added to 2100 ml of water. (Always add acid to water CAREFULLY.) Or, 456 ml of acid per 2 litres of water, at 15°C.

This makes a solution of 1.25 SG, which is suitable for lead accumulators.

Freezing Mixtures

Sodium chloride (salt) 1 part.
Ice 3 parts — reduces temperature 21°C.
Calcium chloride 2 parts — reduces temperature 30°C.
Ice 1 part.

To obtain lower temperatures a mixture of solid carbon dioxide and trichlorethylene may be used (approx. 3/1).

4.13 Conversion Factors

Some Useful Measurements and Conversions

1 gallon = 4.5459 litres.
To convert litres to gallons × 0.220 (22/100).
To convert gallons to litres × 4.544.
To convert litres to US gallons × 0.2642.
To convert pounds to kilograms × 0.454 (9/20).

1 kilogram = 2.205 lbs.
1 litre = 0.219975 gallons.
1 metre = 39.37 inches = 1.09 yards.
10 milligrams = 1 centigram = 0.15432 grains.

1 drop nearly = 1 minim.
60 minims = 1 dram.
1 cc (cubic centimetre) = 0.06103 cu. inches. = 16.894 minims.
1 pint of water = 1.25 lbs.
One micron = 1μ (mu) = 1/1000 of a millimetre = 0.001 mm = 1/2500 inches.
One angstrom unit = 1 ÅU = 1×10^{-7} millimetres.
1 micro-millimetre = $1 \mu\mu$ = 10 ÅU.

SECTION 5

5.1 Safety and Industrial Hygiene

Introduction

In chemical laboratories hazards may arise from the noxious effects of chemicals or from fire or explosions.

Hazards arising from micro-organisms, or radioactive chemicals, require special considerations. General rules and safety precautions are applicable to most chemical laboratories and a list of these should be posted in a prominent position in the laboratory. The attention of newcomers should be drawn to these rules.

The following rules may be found suitable for ordinary chemical laboratories, particularly where large amounts of organic chemicals are handled.

General Safety Rules and Precautions in the Laboratory

1 — Laboratory coats should always be worn in the laboratory for the protection of clothing.

2 — Special care should be exercised when *handling strong acids*, particularly *fuming acids*, *alkalis, bromine, hydrofluoric acid* or other powerful corrosives, *peroxides* and *perchloric acids, chlorosulphonic acids, per-sulphates*.

Winchester-size bottles should be transported by acid carrier. Anti-splash goggles should be used when pouring out these acids. No dangerous acids should be placed on shelves above eye level.

3 — Particular care should be exercised in desposing of waste and reaction residues. All *sodium residues* (even when apparently inactive) should be treated with alcohol before disposal. Empty bottles which may have contained sodium or substances forming explosive compounds should not be left in cupboards or on shelves, but should be disposed of in the appropriate manner.

Potassium may be rendered inactive by 5% alcohol in liquid paraffin. All containers of materials that are not being used immediately and especially those which are to be preserved must be *adequately* labelled.

4 — Protective goggles should be worn on all occasions where hazards may arise, especially when pouring concentrated acids from larger bottles or containers.

A number of other organic solvents can cause eye damage if splashed into the eye, e.g. benzoyl chloride, dimethyl sulphate, chloroform, etc. The vapour of the two former can also be dangerous. Protective eye shields, or goggles should always be worn when pouring these and other hazardous liquids. The transfer of such liquids should also be done in a fume cupboard. Strong caustic solutions are also in this category.

Opening glass phials should be done with care. Read the instructions on the phial and *cool if necessary* first. When opening a phial use a rag or towel to hold it and always away from the face.

Eye injuries, whether from chemical or mechanical causes, must always be considered serious. The best treatment for chemical injury to the eye is immediate and prolonged flushing with water. Obtain medical advice for any eye injury as soon as possible.

5 — Operations involving noxious fumes should be carried out in

fume cupboards. It is worth bearing in mind that although all chemical substances are quite safe if handled with suitable precautions, many common substances present a high health hazard and the toxic effect may be cumulative. Do not allow any chemical to come into contact with the skin and do not breathe any vapour, even if it has a pleasant smell, unless you are quite sure of the absence of an unpleasant physiological effect.

6 — Inflammable liquids should not be handled in open vessels, and should not be brought into the vicinity of naked flames except in small quantities (a few ml).

Inflammable solvents such as alcohol, ether or light petroleum should be distilled on a steam bath or heating mantle.

As a general precaution smoking should not be permitted in the laboratory.

When a person's clothing catches fire *smother the flames quickly.* Use a fire blanket or a laboratory coat if it is handy but do not delay. If near a shower, roll him under and then turn on the water but never let him stand even if you have to be forceful. This procedure prevents injury to the respiratory passages and eyes by the flames which would naturally rise and envelope the head. Never use an extinguisher of any type on a person. The soda-acid extinguisher may damage the eyes whilst the carbon dioxide type may cause severe frostbite.

7 — Handling of glassware — the following rules should be observed:

(a) Fire polish ends of all glass tubing.

(b) Before inserting glass tubing into corks, rubber tubing or stoppers, make sure that the hole is large enough and moisten the tubing or stopper with water or glycerine. Hold the stopper or rubber tubing between thumb and forefinger and not in the palm of the hand.

(c) Do not try to force an oversized stopper into a flask. A cork may be made smaller and softer by rolling it.

8 — When electrical repairs are necessary they should be reported to the responsible person.

9 — Food should never be eaten in the laboratory.

10 — Enter all injuries in a report book. This should be strictly adhered to.

A number of poisons (mercury, hydrogen sulphide, radioactive chemicals) may give rise to hazards. The danger from radioactive chemicals is most likely to arise in laboratories used for casual radioactive work rather than laboratories used specially for radioactivity.

Heating certain chemicals can cause hazardous explosions, particularly the following: hydrogen peroxide, potassium and sodium, phosphorus, per-acids, chlorates and powdered metals. The use of ammoniacal silver nitrate, in most organic-chemistry laboratories, has now been replaced by using Tollen's reagent A & B. This ensures that only small amounts are made up, as required.

Special precautions are necessary to protect the following from areas where flames or sparks may be present: hydrogen, ethers, petroleum solvents, etc. Dangerous explosive mixtures are formed by these and air, and special storage facilities are usually provided. Liquid air will cause organic material to burn explosively. Usually liquid nitrogen is recommended for low temperatures.

Other general considerations are: labelling of containers, storage of gas cylinders — as regards prevention of leaks and observance of the colour code.

Protection from moving parts of machinery is usually included in legislation covered by the Factories Act, but in small laboratory workshops can easily be overlooked.

Electrical appliances, such as water baths with electrical heating elements and thermostats, should be given careful attention to ensure that they are in order. Any leaks or defects should be immediately attended to or the item should be taken out of service.

Dangers arising from new and experimental apparatus are also factors to be considered, particularly with regard to procedures using high-voltage apparatus (e.g. high-voltage electrophoresis).

In biological and bacteriological laboratories it is necessary for special measures and designs to be incorporated in the working area. Most large bacteriological laboratories have air conditioning (by plenum and exhaust methods). Double doors for entry to the laboratories and, in tropical and sub-tropical areas, double fly-proof doors. A special fume cupboard with ultraviolet lighting or a transportation box may be provided for transfer of bacteria. The inside of this box can be sterilised by blowing in steam; the box is portable and ordinary (size 8.5). Surgeons' gloves may be used; the 'wrist ring' of the glove fits round a projection on the box to form a seal. Otherwise a conventional glove box may be used but most of these are not of an easily portable kind. Dangers may arise from airborne microbes or contact and may require special precautions when working with unidentified microbes. The portable box is very useful for this reason and can be used in conjunction with a larger fume cupboard.

Many of the factors applicable to radiological safety may also be applied to bacteriological laboratories, in some cases with less stringency.

Legislation, such as the current 'Health & Safety at Work Act' and 'Dangerous Microbes Act' or other relevant acts, should be displayed prominently in laboratories and references drawn to such legal requirements.